A Programmed Approach to Human Genetics

A Programmed Approach to Human Genetics

ALLEN VEGOTSKY
CYNTHIA A. WHITE
Wells College Aurora, New York

Library of Congress Cataloging in Publication Data:

Vegotsky, Allen, 1931-
A Programmed Approach to Human Genetics

Bibliography: p.
1. Human genetics–Programmed instruction.
I. Stone, Cynthia White, joint author. II. Title.
[DNLM: 1. Genetics, Human–Programmed texts. QH431 V424p 1974]

QH431.V36 573.2'1'077 73-22395
ISBN 0-471-90491-0

Printed in the United States of America

10 9 8 7 6 5 4 3 2 1

Preface

This book, although short, has matured slowly. It began almost a decade ago when the idea of a programmed approach to the genetics of man occurred to one of us (A. V.). For several years, the idea remained dormant. In the spring of 1968, the other of us (C. A. W.), then an undergraduate student, accepted the challenge and began the actual writing of the book. It was nursed along by our continued fascination with human genetics and by constant testing and criticism by undergraduate students at Wheaton College and at Wells College. A final impetus for completion of this book was provided by a grant from the College Center of the Finger Lakes and the offer of secretarial assistance from Wells College.

The book will be useful to undergraduates as a supplement to their course in general genetics. Although there are excellent genetics textbooks for the liberal arts student, most of these tend to have little coverage of human heredity, the topic with the greatest appeal to many students. On the other hand, human or medical genetics textbooks have a too-specialized slant for the undergraduate student who needs an awareness of basic genetic principles, many of which were derived from studies of organisms other than man (for example, microbes, plants, and fruit flies). Thus this programmed text is perhaps best suited as a supplement to a more traditional textbook and has these added advantages:

1. With a programmed text, the student may proceed at his own speed. In our experience, this feature has had a special fascination for some of the brighter students who have been stimulated to go well beyond their assignments.
2. A programmed text makes little demand on faculty time.
3. The emphasis on the genetics of man provides a combination of review as well as new and relevant information.

4. This book emphasizes applications of genetics to other fields (that is, psychology, and medicine, for instance).

We hope that this book will have other uses. As a supplement to texts on human or medical genetics, this book allows the reader to test his comprehension of basic principles. The programmed text may also be helpful to busy professionals who lack course work in genetics and find that an understanding of human genetics is essential. This group will include physicians, biomedical engineers, anthropologists, and psychologists. These readers should complete the programmed text and continue with a more thorough textbook for an indepth introduction to the genetics of man. A listing of such Textbooks of this kind is provided in the bibliography after Chapter 11. Readers who have had a general biology course and who have been exposed to a few statistical tools will appreciate the logical development of this book. However, it is possible to master the concepts discussed here without a formal college course in general biology or statistics.

Almost no attempt is made to deal with the molecular or cellular aspects of genetics—subjects of considerable importance to geneticists. These areas are best treated in a traditional-style text. We feel that the subjects emphasized here (sometimes called Mendelian or classical genetics) are especially well suited to the programmed approach because the concepts lead logically from the simple to the complex.

We are grateful for the careful reviews and suggestions from scientific colleagues Robert E. Hannemann, George A. Hudock, Victor A. McKusick, and Edward R. Pierce as well as valuable guidance from John Wiley's editorial staff, especially Irene Brownstone who reviewed each Frame carefully and improved many of them. The authors are responsible for errors and for the further evolution of our text and appreciate reader comments to aid this process.

Allen Vegotsky
Cynthia A. White

Instructions to the Student and Objectives

This programmed book is written in short segments called frames. Each frame presents some new information and includes a question to answer or a problem to solve so that you can actively apply what you learn as you go along. The correct answer or solution is given below the dotted line. As you read through the book, use a piece of paper or an index card to cover the answers. Slide the card or paper down to the dotted line. Next read the frame and write down your answer. Then check your answer against the one given in the book. If your answer does not agree with the book's answer, be sure you understand why before you go on. Review earlier frames until you understand the problem and the solution clearly.

Readers with a background in genetics may find the first few chapters deceptively easy. However, a review and mastery of the language of genetics will facilitate their understanding of concepts in later chapters. After completing this book the reader should be able to:

- Understand and use the basic terms and concepts of Mendelian genetics as they relate to man.
- Infer genetic relations from pedigree charts.
- Make predictions about families and future offspring from genetic data.
- Begin to approach problems dealing with inheritance like a professional geneticist.

A. V.
C. A. W.

Contents

A Programmed Approach
to
Human Genetics

CHAPTER 1

The Physical Basis of Heredity

If we were to observe a random group of people, we would notice differences in eye color, hair color, and body structure, for example. Yet, if we were to assemble a large family, we would, of course, be impressed by their relative similarity of appearance. This similarity may be explained and predicted by the principles of the science of genetics. It is evident that not all aspects of a family's appearance is inheritable; for instance, although voice range and hair color are inherited (i.e., genetically determined or gene controlled), dialect and hair style are not. The geneticist uses the term "characteristic" for the components of our makeup that are inheritable.

1. Eye color, handedness, ear shape, nose shape, and the like, exemplify genetic characteristics while beard length, temperament, and habits do not. Which of the following would you classify as characteristics: blonde hair, attached ear lobes, thrift, "good taste," an aquiline nose, religious zeal.

Blonde hair (if natural), attached ear lobes, and an aquiline nose are inheritable characteristics while thrift, "good taste," and religious zeal are not characteristics but result from training and environment. The determination as to whether something is or is not a characteristic (i.e., inheritable) is not always easy and is the subject matter of Chapter 10. At this time, however, you need only have an idea of what a characteristic means.

2. The fact that a child possesses the characteristics of his father, mother, or both leads to the conclusion that characteristics are passed from parents to their ______________.

---

children or offspring

3. Characteristics are transmitted from generation to generation by means of information contained in hereditary units called genes. The genes are located on microscopic rodlike structures (chromosomes), which are found in the cell nuclei. With the limited information provided up to this point explain, in your own words, how a child "gets blue eyes" from a parent.

A child must inherit genes from one or both parents conveying information for blue eye color. The genes for blue eyes are contained in chromosomes located in the cell nuclei.

4. There are generally two genes aligned on the chromosomes for every characteristic which an individual possesses.[1] Explain how a characteristic like a receding chin is inherited.

A child must inherit *two* genes from his parents that provide information or specify his receding chin. Again, we might add that the genes are found in the chromosomes of the cell nuclei.

5. The only opportunity during which the hereditary units or genes can be passed from one generation to the next is during sexual reproduction. Since the genes are located or "carried" on the chromosomes, the distribution of genes from parent to child is based on the distribution of

chromosomes. Now, with this new information, modify your explanation on the inheritance of receding chin (previous question).

The child inherits two genes from his parents that designate receding chin. This transfer of information occurs during sexual reproduction.

6. Within each individual, there are 23 pairs or #_____ chromosomes in a typical body cell.[2] This is said to be the "diploid" number for man.

46

7. In reproduction, the egg of the female is fertilized by the sperm of the male. The resulting cell, a "zygote," has the normal or diploid complement of #____ chromosomes.

46

8. The gametes (i.e., egg and sperm) each contribute equally to the chromosome number of the zygote and, therefore, each gamete must contain half the normal complement of chromosomes. This is called the "haploid" number. If, on the other hand, the gametes each had 46 chromosomes their union would result in #___ pairs of chromosomes or #___ total chromosomes.

46
92

9. That is twice the normal number of chromosomes. There is, therefore, a process by which the number of chromosomes is reduced from #___ to #___ in the gametes. In other words, the cell goes from a diploid state to a __________ one.

46
23
haploid

10. Which of the following cells are diploid and haploid? egg cell, zygote, liver cell, sperm

---

diploid: zygote, liver cell
haploid: egg cell, sperm

11. As a result of the reduction in the number of chromosomes in the gametes, the number of genes for each specific characteristic is reduced from two to one (in each gamete). Now, once again apply this new information to your explanation of the inheritance of receding chin.

The parents each possess two genes with information regarding chin shape. When the parents produce gametes, the number of genes for chin shape is reduced to one (per sperm or egg cell). The zygote that forms in reproduction would again have two genes that somehow specify receding chin. (It should be clear to the reader that we are trying to evolve or construct a picture of how inheritance works. We recognize that for the student with recent background in genetics, much of this may appear very elementary; such students should feel free to skim quickly through these questions and answers in this first chapter.)

12. Each sperm or egg contains #___gene(s) that control(s) a particular trait.

---

one

13. The process by which gametes with a haploid number or #____chromosomes are formed is called meiosis.

23

14. The end products in meiosis are gametes containing haploid nuclei. In other words, in the process of meiosis, we go from a diploid cell with two genes specifying chin shape to either sperm or _____ cells containing #___gene(s) for chin shape. However, there are differences in the processes of sperm and egg production.

---

egg
one

15. In spermatogenesis (the production of sperm cells), the end products of the process of ________ are mature functional sperm (haploid). In spermatogenesis, there are two meiotic cell divisions. How many sperm must result? #_____.

---

meiosis
four

16. In oogenesis, or egg production, there are #_____ cells formed as in spermatogenesis. However, in oogenesis, unequal divisions of the cytoplasm occur resulting in three small nonfunctional cells (i.e., polar bodies) and only #_____ functional egg(s). By "functional," we mean a cell that can be fertilized by a sperm to form a zygote that can eventually become a human being.

---

four
one

17. A diploid zygote is formed during fertilization by the fusion of the (haploid, diploid) sperm and the (haploid, diploid) egg.

---

haploid
haploid

18. In the formation of a (haploid, diploid) zygote, each parent contributes a gamete containing an (equal, unequal) number of chromosomes and genes.

---

diploid
equal

19. Thus, from each parent comes # ____ gene(s) for a specific characteristic. The zygote, then, contains # ____ gene(s) for each characteristic.

---

one
two

20. In order to maintain the diploid number of chromosomes while proceeding from a single cell (the zygote) to a multicellular organism, what must happen to the genes and chromosomes during cell divisions?

Prior to each doubling of the number of cells, there must first be a process of replication of chromosomes followed by distribution of half of the chromosomes to one cell and half to another.

21. This process of gene duplication and cell divisions that leads from the zygote to the ____________ is called mitosis. Mitosis is an exact copying process in contrast with __________, which leads to haploid gametes.

---

individual (or organism)
meiosis

22. The end products of mitosis are # ____ daughter cells that are exact duplicates of the parent cell.

---

two

23. The daughter cells have #_____ chromosomes (total) and #_____ genes for each characteristic. In this way, the organism maintains its genetic makeup (information) and chromosome number throughout life.

- -

46
two

BRIEF REVIEW OF CHAPTER 1

24. What is the hereditary link between generations?

- -

the gene or chromosome

25. A child carries genes similar to those of his _________ and his _______. The presence of both types of genes may be traced back to the fusion of __________ and __________ to form a diploid __________.

- -

mother
father
egg
sperm
zygote

26. The sperm and egg, having only one gene per trait, are __________. The zygote is __________.

- -

haploid
diploid

27. In contrast to mitosis, the reduction in number in genes from the normal adult complement to the haploid gamete is accomplished through the process of __________.

- -

meiosis

28. During cell division, the diploid number of chromosomes and genes is maintained by the process of ______________.

---

mitosis

29. One last time; explain what happens to the gene or genes for receding chin in inheritance. Refer to numbers of genes involved, meiosis, formation of gametes, fertilization, development of a multicellular organism, and so on. If you can answer this one question thoroughly, you've really mastered the chapter.

To begin, gametes are formed by two meiotic divisions. The resulting sperm and egg cells are haploid and contain only one gene carrying information regarding chin shape. When the gametes of two parents fuse (i.e., fertilization), a diploid zygote results containing two genes for chin shape. The zygote becomes a multicellular human being as the result of mitotic divisions. In each mitotic division, first there is a duplication of chromosomes (and the genes they contain) and then the chromosomes are partitioned equally into two cells. As a result, we continue to have two genes for chin shape in each new cell. How these genes specify chin shape (or eye color or attached ear lobe) is the subject of the rest of this book.

NOTES

[1] The assumption that each characteristic is determined by a single pair of genes will be reconsidered in a later chapter. Also, it will be explained that each member of the gene pair that determines a characteristic is located on a separate chromosome.

[2] In this book, the symbol # will signify that a *number* is called for in the answer.

CHAPTER 2

Matings Involving a Single Characteristic

Knowing that two genes are present for each trait that a person carries, how can we explain the observation that two brown-eyed parents may have a blue-eyed child? The fact is that those characteristics that we observe in an individual give a clue to his genetic makeup but do not disclose the entire story. If we know what genes for eye color the brown-eyed parents had, then we could make a valid prediction as to the possible eye color or colors of their children. Observable characteristics such as eye color or nose shapes are called "phenotypes;" the genes that specify characteristics constitute the "genotype." We can see a phenotype and, of course, we cannot see a genotype. It is the genotype that determines the nature of the offspring. This chapter provides the reader with the ability to use the methods of the geneticist to deduce genotypes from phenotypes and make predictions about the inheritance of phenotypes.

1. If we assume that handedness is inheritable, and if we are told that two right handed parents have a child, can we conclude that the child will also be right handed?

No. It is the genotypes of the parents (not their phenotypes) that determines whether the child is right handed or left handed.

2. Handedness, hair color, and ear shape are examples of (phenotypes, genotypes).

---

phenotypes

3. An individual who has a widow's peak (a pointed frontal hairline) may carry either (a) one gene for widow's peak and one for straight hair line or (b) two genes for widow's peak. Thus, from the observable (phenotype, genotype) one cannot necessarily specify the (genotype, phenotype) .

---

phenotype
genotype

4. The chromosomes of an individual vary in size and shape (i.e., they are morphologically distinctive). It will be recalled that man has 23 *pairs* of chromosomes. One member of each pair of chromosomes is derived from the gametes of one parent, the other member of the pair is derived from the other parent. Each pair is said to be "homologous" because they are morphologically alike and because they carry genes that determine the same phenotypes. If we looked at a microscope slide of stained chromosomes and found that two chromosomes were alike and larger than the other 44 chromosomes, what could we infer? ____________________

The two larger chromosomes are apparently homologous. One is derived from the mother (i.e., from one of her eggs) and the "homolog" is derived from the father (i.e., one of his sperm cells). The two homologs contain genes that are related to the same phenotypes (and characteristics).

5. Genes that occupy the same place or locus on homologous chromosomes are called "alleles." Two alleles may be identical or they may specify contrasting characteristics. Therefore, in the case of the widow's peak (see Question 3*a*), the individual contains #____ alleles, one for __________ __________ and one for ______________________________.

---

two
widow's peak
lack of widow's peak

6. An allele that shows as an observable characteristic in the presence of a contrasting allele is called "dominant." The contrasting allele that is *not* shown in the phenotype is said to be "recessive." The allele for widow's peak is ____________________. The allele for straight hair line is ____________________.

---

dominant
recessive

7. The dominant allele is generally denoted by a capital letter, the recessive allele with a small letter. Thus a person carrying one dominant allele for widow's peak and a recessive allele for straight hair line would have a genotype of Ww. "Darwin's ear point" is a phenotype consisting of a tubercle or point on the infolded edge of the outer part of the ear. The allele for Darwin's ear point is dominant over the allele for no ear point.
Write the genotype for:

(a) An individual carrying two alleles for Darwin's ear point __________.
(b) An individual with one allele for Darwin's ear point and one for absence of ear point __________.
(c) An individual carrying two alleles for absence of ear point __________.

DD
Dd
dd

8. If the two alleles are the same, then the individual is said to be homozygous for that trait. On the other hand, a heterozygous individual has two different alleles for a trait. (In this book, the term "trait" will be used synonymously with "characteristic." Technically, a trait is a more general descriptive term. Thus, red hair is a characteristic and hair color is a trait.) Which of the following genotypes are heterozygous and homozygous?
Dd, WW, dd, Aa, Qq

---

heterozygous: Dd, Aa, Qq
homozygous: WW, dd

9. We have seen that Darwin's ear point and widow's peak are governed by dominant alleles. A few other examples of dominant alleles are: (1) production of skin pigment, (2) free (or hanging) ear lobes, (3) straight-tipped nose, (4) polydactyly (presence of more than 10 fingers or 10 toes), and (5) Huntington's chorea (a disease of the nervous system). The corresponding recessive alleles are (1) lack of production of skin pigment, (2) attached ear lobes, (3) upturned nose tip, (4) 10 fingers and 10 toes, (5) lack of Huntington's chorea. Sometimes students assume that dominant traits (i.e., traits determined by dominant alleles) are more common than recessive ones because they are expressed in the heterozygous as well as homozygous dominant individual. Is this assumption valid?

- -

No. As we note above there are dominant traits such as polydactyly and Huntington's chorea that are rare in the population.

10.[1] If a father is genotypically QQ, he will produce sperm in the process of meiosis that are Q in genotype.
In this process we have gone from the diploid state to the __________ state.

- -

haploid

11. Similarly, if a mother is QQ all the eggs she produces will be genotypically __________.

- -

Q

12. If a sperm cell (haploid) of genotype Q is fused with an egg cell (haploid) of the same genotype, the zygote would be a __________ cell genotypically __________.

- -

diploid
QQ (or homozygous)

13. Therefore, when both parents carry a dominant allele and are homozygous for that allele, their offspring must carry the ______________ allele and must be ______________ for that allele.

---

dominant
homozygous

14. The offspring in Question 13 are both phenotypically and ______________ like their parents.

---

genotypically

15. Now consider the case in which the father carries the dominant allele (Q) and is homozygous for that allele and the mother carries the recessive allele for the same trait and is homozygous. The father, therefore, is genotypically ______________ and the mother is genotypically ________.

QQ
qq

16. The sperm resulting from meiosis in the father will be genotypically ______ and the eggs similarly produced by the mother will be genotypically ______.

Q
q

17. When the egg is fertilized by the sperm, the genotype of the zygote would be ______. Thus, the genotype of the offspring is determined by the genotypes of the parental gametes.

Qq

18. Which parent will the offspring resemble phenotypically? ________. Which parent will the offspring resemble genotypically? ____________________.

---

the father (because Q is dominant)
neither (because the offspring is Qq, while the parents are QQ and qq)

19. The Punnett square or checkerboard is a method of predicting the ratios of the genotypes of offspring given the genotypes of the gametes. From these genotypes, the phenotypes may be inferred.
Let W = widow's peak[2]
Let w = straight frontal hairline
Assume that two parents are heterozygous for hair form. Then each sperm would carry either ____ or ____. Each egg would carry ____ or ____.

W
w
W
w

20. A Punnett square is shown below. The top line shows the alleles that the sperm may carry. The alleles that the egg may carry are placed vertically. In center boxes are placed the various combination of alleles that may arise in the zygote by random fertilization. Summation of the boxes gives the ratios of genotypes from which the phenotypes may be inferred. Fill in the empty boxes.

♀ \ ♂	W	w
W		
w		

♀ \ ♂	W	w
W	WW	Ww
w	Ww	ww

21. From the Punnett square in the previous question, list the genotypes that result in a mating of Ww x Ww. ______________________________
____.

WW
Ww (twice)
ww
hence a ratio of 1:2:1

22. The genotype, Ww, is heterozygous and also called a monohybrid. (The term "monohybrid" is more commonly used in describing the genetics of organisms other than man.) A mating such as Ww x Ww is, therefore, called a monohybrid cross. List the phenotypes of the offspring from this monohybrid cross. ______________________________.

---

widow's peak; individuals with a straight hairline

23. Analysis of the Punnett square reveals that the phenotypic ratio of offspring in this cross is _______ to _______. The genotypic ratio is _______ to _______ to _______. You may assume that these ratios apply to all monohybrid crosses. (We will reconsider this assumption in a later chapter.)

3:1
1:2:1

24. Suppose one of the male offspring who is genotypically Ww marries a female who is genotypically ww. The heterozygous male can produce sperm that carry either _____ or _____ in a ratio of 1:1.

W

w

25. The genotype of a straight hair-lined female would be ____. An egg produced by this homozygous recessive female would contain the allele _____.

---

ww

w

26. What are the possible combinations of alleles that the zygote can carry in a cross of Ww x ww? Use the Punnett square method.

♀ \ ♂	W	w
w	Ww	ww

Although the sperm may carry W or w, the female may carry only w. Since we are concerned with ratios, w is written in only one row of the Punnett square rather than the scheme shown below. (Both, however, would give identical ratios.)

♀ \ ♂	W	w
w		
w		

27. The genotypic ratio of a Ww x ww cross is ______ to ________.

---

one Ww to one ww

28. What is the phenotypic ratio? ______________________.

---

one widow's peaked to one nonwidow's peaked

29. The above cross (#26) between one parent who carries only the recessive allele and is homozygous for a trait, and the other parent who is heterozygous for this trait is called a "test cross." The genotypic and phenotypic ratios of a test cross are # _____ to # _____.

---

one
one

BRIEF REVIEW OF CHAPTER 2

30. Let "A" symbolize the dominant allele for achondroplasia (a skeletal disease) and "a" signify the recessive allele for absence of this disorder. Diagram the Punnett squares for all the different possible crosses involving this gene pair. (Hint: there are six possibilities.)

(a) AA x AA

♀ \ ♂	A
A	AA

(b) aa x aa

♀ \ ♂	a
a	aa

(c) AA x aa

♀ \ ♂	A
a	Aa

(d) AA x Aa

♀ \ ♂	A
A	AA
a	Aa

(e) aa x Aa

♀ \ ♂	a
A	Aa
a	aa

(f) Aa x Aa

♀ \ ♂	A	a
A	AA	Aa
a	Aa	aa

31. Which of the above crosses is a monohybrid cross, a test cross, neither?

____

____

---

monohybrid cross: f
test cross: e
neither: all others

32. List the genotypic ratios and phenotypic ratios for each of the above crosses.

CROSS	GENOTYPIC RATIO	PHENOTYPIC RATIO
a^3	–	–
b	–	–
c	–	–
d	1:1	1:1
e	1:1	1:1
f	1:2:1	3:1

33. Give the possible genotypes for each of the following individuals:
(a) A woman who does not have achondroplasia __________.
(b) A child who is heterozygous for the condition__________.
(c) A man who has the disease __________.
(d) An individual who is homozygous recessive __________.

(a) aa
(b) Aa
(c) AA or Aa
(d) aa

NOTES

[1] If you find questions 10 through 12 difficult, review Chapter 1.

[2] This is the geneticist's way of briefly stating the following: let the symbol, W, represent the allele for the production of a "widow's peak" or a pointed frontal hairline.

[3] No ratio is recorded for a, b, and c since only one genotype and one phenotype is found.

CHAPTER 3

The Pedigree Chart in Genetic Analysis

One method for making predictions about offspring is through analysis of pedigree charts. Symbols used in pedigree charts include:

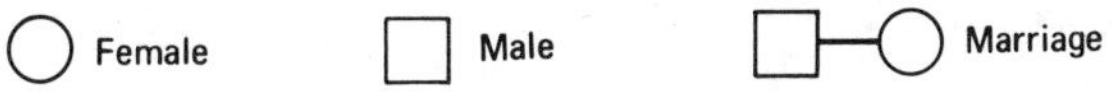

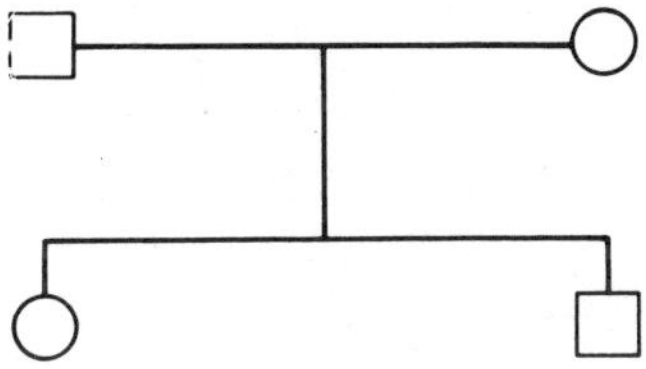

Interpretation: A marriage (or simply mating) of a couple who have two offspring, a daughter, and later a son. We know the daughter was born first because she is shown farthest to the left. The horizontal line connecting the two sibs (or siblings) is called a sibling line.

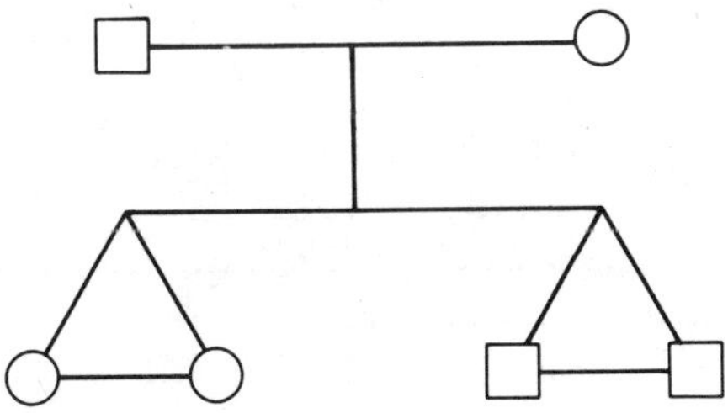

Interpretation: A couple with two sets of monozygotic (i.e., identical) twins. (Monozygotic twins are derived from one sperm and one egg, which is indicated by a horizontal line between twins.)

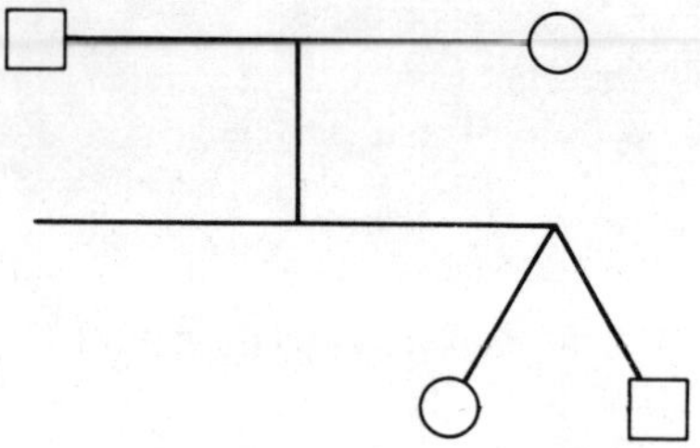

Interpretation: A couple with dizygotic (i.e., fraternal) twins. (Dizygotic twins are derived from separate eggs each of which is fertilized by a separate sperm. Such twins are no more alike than two sibs.)

or

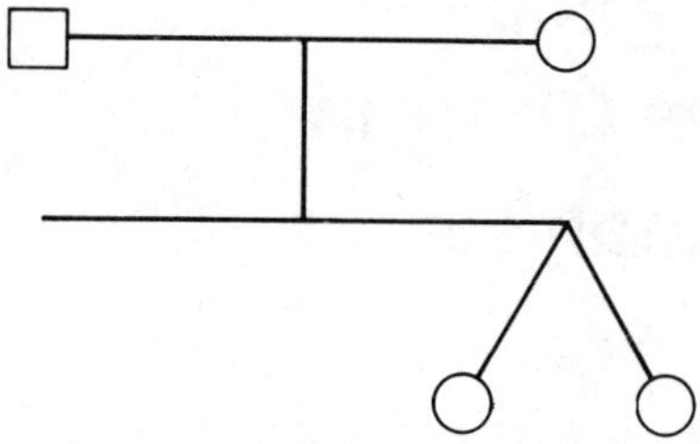

or

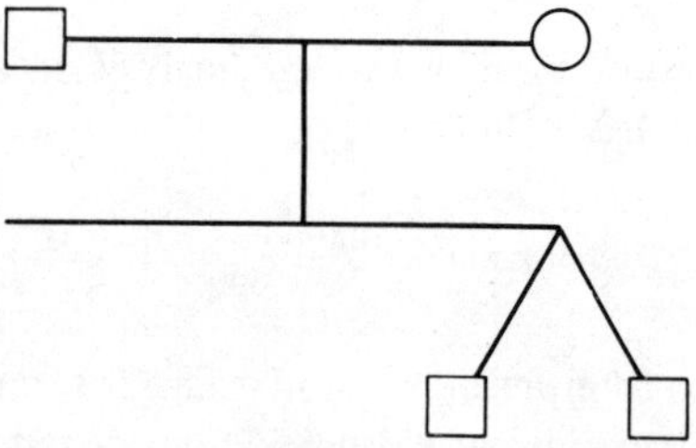

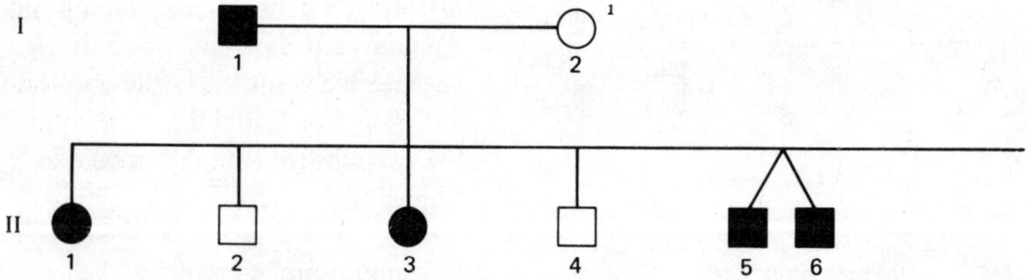

1. Consider the pedigree chart above. What is the relationship between the following pairs? Individuals II-2 and II-3 ________________________ ; II-3 and I-1 ________________________; II-5 and II-6 ________________________.

---

brother and sister (or siblings)
daughter and father
twin brothers; they are dizygous twins since there is no horizontal line connecting the two.

2. In a pedigree chart, shaded squares or circles represent individuals with a particular defined trait. In the chart above, there are # ______ males and # ______ females with the trait and # ______ males and # ______ females that do not have the trait.

- -

three
two
two
one

3. Although some individuals cannot taste phenylthiocarbamide (PTC), others have the ability to recognize a distinct generally bitter taste in this chemical substance. A pedigree chart showing a family of tasters and nontasters is shown below.

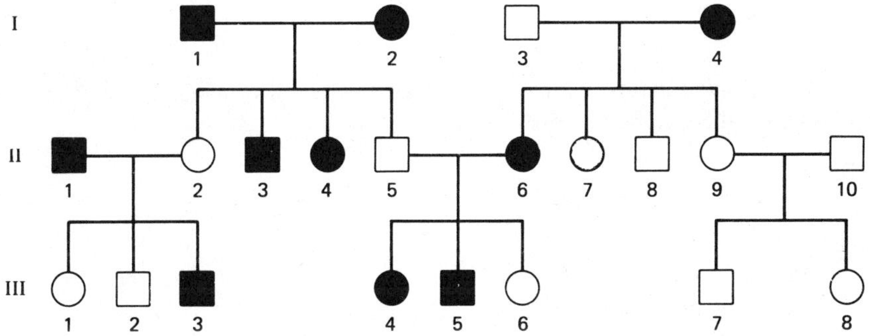

In the chart, the total number of tasters is ____, and the number of non-tasters is ____.

- -

10
12

4. The relationship between individual II-4 and II-5 is ______________ and ______________. The relationship between I-3 and III-6 is that of ______________ and ______________. The female, II-2, is (older/younger) ______________ than II-4. What is the relation between II-1 and II-5? ______________; between III-3 and II-4? ______________; between III-1 and III-5? ______________.

- -

sister and brother
grandfather and granddaughter
older
brothers-in-law
nephew and aunt
first cousins

5. From the pedigree chart shown, it is possible to establish whether the ability to taste PTC is a dominant or recessive trait. This is accomplished by testing the two possible hypotheses for inheriting the trait. (In reality there are more than two possibilities; more complex genetic pedigrees are treated in later chapters.) The allele for tasting PTC must either be ________________ or ________________.

- -

dominant
recessive

6. Take a few minutes to study the pedigree chart following Question 3. Is tasting likely to be a recessive trait?

- -

No. (If the reasons for the conclusion are not clear, study the following arguments especially carefully.)

7. There are several arguments against tasting being a recessive trait. The most obvious argument refuting the hypothesis that tasting is a recessive trait is seen in the mating between individuals #I-1 and #I-2. If the trait were recessive, both parents would have to be genotypically ________________ ________________________.

- -

tt (or homozygous for t). Now, let's consider the offspring in relation to the recessive hypothesis.

8. In that case, all sperm produced by I-1 and eggs produced by I-2 would be genotypically ______ and all offspring would have to be genotypically ________________________________ and phenotypically ____________________. Is this prediction consistent with the chart?

- -

t
tt (If you're not convinced, try a Punnett square)
tasters
no, since II-2 and II-5 are nontasters

9. What does this do to our hypothesis that tasting is a recessive trait?

____

____

____

- -

The observation that two parents who are tasters have one or more children who are nontasters refutes the recessive hypothesis.

10. Is tasting inherited as a dominant trait? ________________
Study the chart to see if this hypothesis is supported.

- -

Yes. The hypothesis that tasting is a dominant trait is valid based on all the data in the chart. The evidence is considered in detail in the next three questions.

11. Theoretically, tasters could be genotypically ________ or ________.

- -

TT
Tt

12. Can you assign genotypes to individuals I-1, I-2, and I-4? Explain.

____

____

____

____

- -

Yes. They must each be Tt. This conclusion is based on their offspring (generation II). If I-1 or I-2 or I-4 were TT, then all their offspring would have to have at least one dominant allele and they would *all* be tasters. This is not the case.

13. II-3 and II-4 may be genotypically ________ or ________.

TT
Tt

Other arguments in support of the dominant hypothesis are the following. (1) There is no way to negate the hypothesis with the data supplied. (2) If the hypothesis is valid, we may infer that I-4, II-1, and II-6 are Tt, and their mates are tt. The results of each of these matings should be a 1:1 ratio (Note: these are test crosses as discussed in Chapter II, Frame 28); in assessing this prediction, we see that of a total of 10 progeny from these three matings, 6 are nontasters and 4 are tasters. Considering the small size of the family, the results are consistent with the 1:1 ratio anticipated.

One form of albinism is related to the lack of melanin pigment in the skin, hair, and eyes. In the following chart, ■ or ● = those having the albino trait.

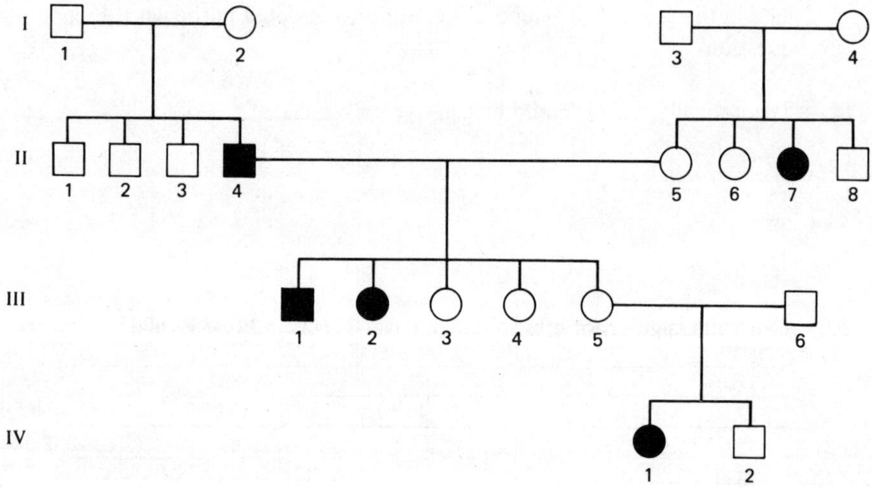

14. We may begin with the assumption that albinism is either a ________ or ________ trait.

---

dominant
recessive

15. If albinism is a dominant trait, is it possible for parents I-3 and I-4 to have a daughter like II-7? ______________

- -

No

16. The inference that albinism cannot be dominant is based on the following lines of reasoning: if we temporarily accept the "dominant" hypothesis, then all darkened squares and circles would be genotypically ________ or __________ and all unshaded squares and circles would be __________.

- -

AA
Aa
aa

17. For example, individuals I-1 and I-2 would be genotypically __________. What would be the genotypes and phenotypes of their offspring?

____

____

- -

aa
genotypically – aa, phenotypically – not albino (this must be true because gametes produced by I-1 and I-2 could only transmit the allele, a.)

18. Is this prediction supported by the pedigree chart? ____________

- -

No

Therefore, albinism is not a ____________ trait and by elimination must be a ____________ trait.

- -

dominant
recessive

19. What other information derived from the pedigree chart supports the hypothesis that albinism is a recessive trait?

---

(1) There is no way to negate the hypothesis.
(2) The crosses of I-1 × I-2, I-3 × I-4, and III-5 × III-6 are each monohybrid crosses. The results of these crosses (seven nonalbino and three albino) is close to the predictable ratio for a cross of Aa × Aa.

BRIEF REVIEW OF CHAPTER 3

20. Deduce and list all individuals that must be heterozygous based on the above pedigree chart. Analyze the pedigree chart supplying the genotypes for the remaining individuals.
Let A = nonalbino (normal pigmentation)
a = albino

I-1	Aa
I-2	Aa
I-3	Aa
I-4	Aa
II-1	Aa or AA
II-2	Aa or AA
II-3	Aa or AA
II-4	aa
II-5	Aa
II-6	AA or Aa
II-7	aa
II-8	AA or Aa

III-1 aa
III-2 aa
III-3 Aa
III-4 Aa
III-5 Aa
III-6 Aa
IV-1 aa
IV-2 Aa or AA

(It is not always possible to establish the exact genotypes of all the offspring. Therefore, in some cases, there are two genotypic possibilities.)
If you had trouble with this answer, consider the following points and then review the chapter. All albinos (shaded squares or circles) must be aa. All others must be AA or Aa. If a nonalbino individual has an albino child, that parent cannot be AA but must be Aa; this explains the genotypes for: I-1, I-2, I-3, I-4, II-5, III-5, and III-6. Also, if a normally pigmented individual has an albino parent, the "normal" individual must receive the allele, a, from the albino parent. This explains the genotypes of III-3 and III-4 and provides an additional argument for the genotype of III-5.

21. This is an optional question. The disease, homocystinuria, is known to be inherited via a recessive gene. Construct a pedigree chart for an imaginary family that proves the recessive hypothesis.

You're on your own since there are an infinite number of answers. If you really want your pedigree chart checked out, send a copy of your chart to the senior author (A. V.).

NOTES

[1] The symbols I and II signify the parental and first filial generation, respectively. The numbers are used for convenience in referring to offspring and also to indicate the birth order of offspring within a given sibship. In other words, male II-3 is older than female II-4.

CHAPTER 4

Matings Involving More Than One Characteristic

Thus far we have dealt with the simplest kind of mating involving one set or pair of alleles. We have seen how by the use of the pedigree chart it is possible to determine whether a trait is dominant or recessive. Next we are going to consider matings involving two or more pairs of alleles. The principles involved are the same as those used with matings involving single alleles since each pair of genes is transmitted independently.

1. Primordial dwarfism is a type of dwarfism in which the body proportions remain normal. In the population, there are ___(few, many)___ individuals who have this trait.

- -

few (just look around)

2. It is known that primordial dwarfism is a recessive trait. Therefore, consider the case of a dwarf male. All of his sperm would carry the ______ for ____________.

- -

allele
dwarfism

3. Distichiasis (two rows of eyelashes) is a common trait and is dominant. A female who is heterozygous for distichiasis may produce eggs carrying the dominant or the recessive allele. A female with two rows of eyelashes produces eggs that carry only the (dominant, recessive) allele if she is homozygous for the distichiasis.

---

dominant

4. If she is of normal height and does not carry a gene for dwarfism, she would be (homozygous, heterozygous) for normal stature; her eggs would carry only the allele for the ______.

---

homozygous
lack of dwarfism

5. A dwarfed individual with a single row of eyelashes would carry # ______ recessive alleles for eyelashes.

---

two

6. Consider a female of average height. Assume she has two rows of eyelashes and that her family history reveals that there were no dwarfs in her family and there were no individuals with a single row of eyelashes.
Let D = distichiasis (i.e., double row)
d = lack of distichiasis (i.e., single row)
N = nondwarfism ("normal")
n = dwarfism
Her genotype is probably ______.

DDNN (We hedge here because it is possible for a recessive allele to be passed along in a family pedigree for many generations without being manifested. This would be especially true for a rare recessive gene like n because it would not be likely for two heterozygous individuals to mate.)

7. A mating between a female genotypically DDNN and a dwarfed male with one row of eyelashes would result in an individual who is genotypically ___________ .

DdNn

8. All offspring from such a cross would have ___________ height and # _______ rows of eyelashes.

---

normal
two

9. This would be true because all zygotes would be formed by fusion of a ______________ sperm with a __________ egg.

---

dn
DN

10. If, instead, the female was genotypically DdNn, her eggs would carry ____ , ______ , _______ , or _______ in a ratio of __:__:__:__.

DN
Dn
dN
dn
1:1:1:1 (We assume that the alleles governing primordial dwarfism and distichiasis are carried on two different chromosomes.)

11. In the case of a DdNn male, his sperm would carry (the same, different) combinations of alleles as in Question 10.

---

the same

12. What genotypes are possible from a mating in which both parents are heterozygous for the two traits, dwarfism and distichiasis? Use the Punnett square method. (It may be noted that both parents are dihybrids and this is a dihybrid cross[1].)

♀ \ ♂	DN	Dn	dN	dn
DN				
Dn				
dN				
dn				

♀ \ ♂	DN	Dn	dN	dn
DN	DDNN	DDNn	DdNN	DdNn
Dn	DDNn	DDnn	DdNn	Ddnn
dN	DdNN	DdNn	ddNN	ddNn
dn	DdNn	Ddnn	ddNn	ddnn

13. The results of the Punnett square do not give the actual number of progeny but, as seen in Chapter 2, they do give a ______________ of genotypes in the offspring. The predictions are unlikely to be accurate if the number of offspring is very ______________ .

---

ratio
(The results are predicted on the assumption that the fertilization of eggs by sperm cells is a random process.)
small

14. Of the 16 possible combinations in the Punnett square, phenotypically: # ______________ are normal height with two rows of eyelashes. # ______________ are normal height with one row of eyelashes. # ______________ are dwarfs with two rows of eyelashes. # ______________ are dwarfs with one row of eyelashes.

---

nine (DDNN, DDNn, DdNN, DdNn)
three (ddNN, ddNn)
three (DDnn, Ddnn)
one (ddnn)

15. How many different genotypic classes result from DdNn x DdNn? # ____

---

nine (This is evident from the Punnett square and shown in answer to Question 14.)

16. Using the Punnett square method, determine the results of a mating of a DdNn individual with a ddnn individual. List the genotypes that can result and their ratios.

♀ \ ♂	DN	Dn	dN	dn
dn	DdNn	Ddnn	DdNn	ddnn

genotypic ratio = DdNn:Ddnn:ddNn:ddnn / 1:1:1:1

17. The case of a mating involving one individual who is homozygous for all traits and the other who is heterozygous for all traits is called a ________ ____________________.

---

test cross (Note the similarity to the test cross in Chapter 2.)

18. The phenotypic ratio found in a monohybrid cross of two heterozygous parents is ________ to ________.

---

three
one

19. In the dihybrid cross covered in Question 14, the phenotypic ratio is ___:___:___:___.

9:3:3:1

In the rest of this chapter, an algebraic approach to genetics is given. These techniques will enable the reader to solve problems that could also be solved by the Punnett square; however, the mathematical method is usually quicker and more convenient. The reader who is shaky with the binomial expansions that follow may wish to refer to an introductory algebra textbook such as Selby's *Practical Algebra*, a Wiley Self-Teaching Guide.

20. Consider the binomial expansion $[(3x) \times (1y)]^2$. If one treats xy as if it were different from yx, then the solution to such a problem is $9x^2 + 3xy + 3yx + 1y^2$. The numerical coefficients from the expansion, (i.e., 9, 3, 3, 1) give the ratio of a ____________________ cross.

---

dihybrid (This is no mere coincidence. Genetic ratios stem from random assortments of genes on chromosomes in meiosis.)

21. If there had been three segregating genes, the phenotypic ratio would be 27:9:9:9:3:3:3:1 or the same as the coefficients of $[(3x) + (1y)]^n$ where $n = \#$ ______? In other words, if n is the number of allele pairs involved

in a cross, the phenotypic ratio of offspring is equal to ____________ raised to the ____________ power.

- -

3
$3x + 1y$
nth

22. In Question 20, the phenotypic ratios were calculated by using the ratio of a ____________ as the base and raising that to a power equal to the number of ____________ involved.

- -

monohybrid cross (i.e., the numerical coefficients 3 and 1)
pairs of alleles

23. The genotypic ratio of a monohybrid cross is ___:___:___.

- -

2:1:1 (or 1:2:1)

24. The genotypic ratio of a dihybrid cross may be obtained by expanding the coefficients from the genotypic ratio of the monohybrid cross (i.e., $[(2x) + (1y) + (1z)]^2$ giving a ratio of 4:2:2:2:2:1:1:1:1). Note: this procedure is analogous to obtaining a phenotypic ratio by expanding the coefficients from the phenotypic ratio of 3:1 as explained above. (In both cases "*xy*" is considered to be distinct from "*yx*.") Thus, the genotypic ratio for a cross where *n* pairs of alleles are being considered is equal to the coefficient of ____________ raised to the ____________ power.

- -

$(2x + 1y + 1z)$
nth

25. Thus far, we have considered the genotypes or phenotypes that result from a particular cross. To find the frequency at which a *particular* genotype appears, we may apply simple probability theory. For example, if there are one white ball and one black ball in a box, the probability of drawing

the white ball on the first try is 1:2. If there are two white balls and one black ball the probability of drawing the white ball is ___ : ___.

2:3

26. The same principle may be used to consider an infinite number of problems in genetics. The frequency of a given phenotype appearing in offspring is given by the fraction t/T where t = number of individuals with the phenotype in discussion and T = total number of combinations in a given mating cross (i.e., the number of boxes in the Punnett square). Thus, T, the total number of combinations in a monohybrid cross is # ______. Similarly, t, the total number of individuals having the most common phenotype for the mating is # ______. Therefore, the frequency of that phenotype occurring is # ______.

4
3
3/4

27. The frequency of appearance of the most common phenotype in a dihybrid cross is # ______.

9/16 (cf. the Punnett square in Question 12)

28. In a monohybrid cross, the number of phenotypes or classes is two, one possessing at least one dominant allele; and one with # ______ ______ alleles.

---

two recessive

29. In a dihybrid cross the number of phenotypic classes is # ______.

---

four (For example, analysis of the Punnett square in Question 12 shows these four phenotypes: (1) distichiasis and normal height, (2) distichiasis and dwarf, (3) single row of eyelashes and normal height, and (4) single row of eyelashes and dwarf.)

30. The frequency of the appearance of a particular genotype is found in the same way as the frequency of a phenotype. (See Question 28.) In a monohybrid cross the most common genotype is ______________ in the offspring.

Aa (or heterozygous)

31. The frequency of Aa appearing in a monohybrid cross is # __________.

2/4 or 1/2

32. The frequency of AaBb appearing in a dihybrid cross is # __________.

4/16 or 1/4 (Refer to the Punnett square in Question 12 and calculate t/T).

NOTES

[1] This is the language of the general geneticist. The specialist in human or medical genetics might substitute the word "mating" for "cross" and the phrase "heterozygous with respect to two pairs of alleles" for "dihybrid." For convenience, the simpler terminology was used.

CHAPTER 5

Modified Genetic Ratios

The ratios discussed in previous questions are generally valid in making genetic predictions. There are, however, certain important exceptions that modify the predictable results. Several of these exceptions are now considered.

1. Sickle-cell anemia has been traced to a recessive gene, s, and to the production of a modified hemoglobin (blood protein). The genotype of a sickle-cell anemic person is ______. A normal individual would be expected to be genotypically ________ or _______ .

---

ss
SS
Ss

2. If this is true, then SS and Ss individuals should be phenotypically (identical, distinct) with respect to this trait.

---

identical

3. In reality, the heterozygous individual and the homozygous dominant individual differ in their hemoglobin content and in other ways. Therefore, the allele for normal hemoglobin is not completely __________ over the allele for sickle-cell anemia. This is called incomplete (or intermediate) dominance.

---

dominant

4. Assume that the gene for tallness is incompletely dominant over the gene for shortness. (The inheritance of height is actually more complex than this model suggests.) If the mating Tt x Tt were studied, the genotypes of offspring would be ________ , ________ , and ________ .

TT
Tt
tt

5. The genotypes are (identical to, different from) those expected if T were completely dominant.

---

identical to

6. The phenotypes are (identical to, different from) those expected if T were dominant. Tt, where there is incomplete dominance, represents an individual who is (tall, short, medium) .

---

different from
medium

7. Note that in cases of incomplete dominance, the phenotypic and genotypic ratios are (the same, different) .

---

the same (If you are not convinced, try a Punnett square)

8. In comparing complete and incomplete dominance, the phenotypic ratio of a monohybrid cross shifts from ___:___to ___:___:___.

3:1
1:2:1

9. A number of inheritable diseases manifest incomplete dominance. An example is Friedreich's ataxia. Homozygous recessive individuals show poor muscular coordination. What might we predict for heterozygous individuals? ____

____

- -

Heterozygotes should differ from normal (homozygous dominant) individuals. The former have been found to have excessive curvature of the sole of the foot and other indications of poor muscular coordination.

10. Another type of gene interaction that modifies normal genetic ratios is called epistasis. When one gene masks the effect of another nonallelic gene, epistasis exists. To give a crude and farfetched example, assume there was one gene for formation of arms and another for formation of fingers, then if no arms were formed, no fingers could be seen.
Let A symbolize a dominant gene for absence of arms and F for a dominant gene for finger formation. Give the phenotypes for the following individuals: (1) AAff ______________________, (2) Aaff ______________________, (3) aaff ______________________.
An individual who is genotypically AAFF (would, would not) have arms and (would, would not) have fingers. This last case illustrates the principle of______________________.

- -

(1) no arms, no fingers
(2) no arms, no fingers
(3) arms, no fingers
would not
would not
(In spite of the F gene for finger formation, if a person has no arms, he can't possibly have fingers.)
epistasis

11. If an AaFf individual were mated with a person of identical genotype, would we expect the typically phenotype 9:3:3:1 ratios predicted for a

dihybrid cross? ______________________________

____

No, because the genes are not acting independently of each other; some combinations will have genes that are masked by other genes.

12. What phenotypes would be present from AaFf x AaFf?

Hint: use the Punnett square.

---

12 armless
3 with both arms and fingers
1 with arms and no fingers

In epistasis, we see that the phenotypic ratio of a dihybrid cross is changed from 9:3:3:1 to 12:3:1. How about the genotypic ratio? ____________

____

---

no change (compare your Punnett square with the one in Chapter 4, Question 12).

13. In the above matings, we have seen that a dominant gene can be epistatic to another (nonallelic) pair of genes. Similarly, it is possible to have recessive genes mask nonallelic genes. For example, as we have already seen, albinism is a recessive trait. There are several additional pairs of genes (i.e., P_1, P_2, P_3[1]) that control production of pigment but the recessive genes for albinism (i.e., aa) are epistatic to the pigment genes. What would

be the phenotype for a person who is $aaP_1P_1P_2P_2P_3P_3$? ________
Is is possible for two black parents to have an albino child?

____

- -

albino
yes, if they each carry one gene for albinism.

NOTES

[1] The question of multiple gene inheritance will be treated in more depth in Chapter 9.

CHAPTER 6

The Determination of Sex

In general, one of the most obvious phenotypic traits of an individual is sex. We recognize in an intuitive way that an infant has an equal chance of being male or female.[1] It will soon be made evident that the 1:1 ratio has a genetic basis. A clue to the basis of inheritance of sex came from careful studies of microscope slides of chromosomes. Such studies revealed a subtle yet dramatic difference between the chromosome complex of males and females. The significance of this finding will be examined in this chapter.

1. In a large human family, we would predict that about ____% of the offspring would be male and ____ % would be female.

50
50

2. Since the 1:1 ratio of births of males and females is attributed to chromosomes, the chromosomes of the two sexes must be (identical, different) .

---

different

3. In stained cells, the 23 pairs of human chromosomes may be seen. It is possible for an experienced cytogeneticist to group the chromosomes according to size and shape (i.e., morphologically). Sometimes, this is

done by actually cutting out the outlines of chromosomes from an enlarged photograph. The sequential arrangement of chromosomes from the largest pairs to the smallest pair is called a karyotype. What can we say about the karyotypes of the two sexes based on this chapter thus far?____________

____

____

- -

At this point, we can state that the karyotypes of the two sexes must differ since karyotypes are pictures of chromosomes and the chromosomes of the two sexes differ.

4. The last or twenty-third pair of chromosomes in a karyotype is exceptional. These two chromosomes are correlated with the sex of an individual and, accordingly, are called sex chromosomes. Unlike this pair of chromosomes, the other 22 pairs of chromosomes are called autosomes and are similar in appearance in the two sexes. Whether an individual is male or female depends on the nature of the # _____ sex chromosomes. Therefore, the sex chromosomes must be (the same, different) in the male and female. Each of the other 22 pairs of chromosomes (called ________________), is similar in size and shape in the two sexes.

- -

two
different
autosomes

5. In the human female, the two sex chromosomes are similar in appearance, relatively large, and are designated as "X" chromosomes. In the male, however, the sex chromosomes include one "X" chromosome and one very small chromosome termed a "Y" chromosome. Thus, an individual with two X chromosomes must be a ____________while an individual with an X and a Y chromosome must be a _______________ .

- -

female
male

6. In other words, it would seem that the chromosomal basis for an individual to be male is possession of one ____chromosome and one ____chromosome, whereas the female would result from two ___chromosomes. This would

suggest two alternative explanations for maleness. Either an individual is a male because he has only one X chromosome or because he possesses a Y chromosome. Conversely, two different explanations for femaleness could be hypothesized. What are they? ______________________________

____

____

____

X
Y
X
An individual is a female (1) because she has more than one X chromosome, or (2) she is a female because she lacks a Y chromosome.

7. The choice of the correct basis for chromosomal determination of sex may be made from studies of exceptional individuals. For example, karyotypes with two X and one Y chromosomes have been found. What may be inferred from the fact that such XXY individuals are always male?

____

____

Similarly, karyotypes with one X chromosome and no Y chromosome are known; such X humans are invariably female. Is this observation consistent with the inference drawn from the discovery that XXY humans are males?

____

The presence of a Y chromosome (and not the number of X chromosomes) determines maleness.
Yes, both observations suggest that human sex is determined by the presence or absence of the *Y* chromosome.

8. Cytogenetic and genetic studies have revealed that the 46 human chromosomes exist as pairs. Each pair is morphologically similar and deals with the same genetic traits. These pairs are termed "homologs." If a chromosome carries an allele for eye color, then its homolog must carry an allele for a (similar, different) trait.

---

similar. If the reader has difficulty with this reasoning, he should review the first five frames of Chapter 2 and then begin this chapter over again.

9. In the human female, there are 23 pairs of homologous chromosomes. Is this also true for the human male? ____________ Explain. ________
____
____
____
____

---

no
The male sex chromosomes are not homologous. Recall that the X and Y chromosomes are different in size and structure. We will see in the following chapter that these two chromosomes are genetically very different.

10. We have seen before that, in meiosis, we begin with diploid cells containing 23 pairs of chromosomes and produce (haploid, diploid) gametes with #______ chromosomes.

---

haploid
23
(If necessary, review Chapter 1.)

11. In each segregating pair of homologous chromosomes, one chromosome will go into one gamete and the other ____________________ will go into another ______________ .

---

homolog (or chromosome)
gamete

12. Thus, the 23 chromosomes that go into each gamete are distinct from one another. Can any two of these 23 chromosomes be homologs? ________
____
____

No. In normal meiosis, each pair of homologous chromosomes is divided into two different cells. (It is true that exceptions can occur; on rare occasions, mistakes are made in a process of nondisjunction. This subject is treated in more depth in some of the texts listed in the bibliography.)

13. In gametogenesis, the female will produce egg cells containing # ______ X chromosome(s) and the male will produce sperm cells carrying either an ____ chromosome or a ____ chromosome.

- -

one
X
Y

14. In the process of fertilization, there must be a fusion of two haploid __________ to form a _______ zygote. The sex of the individual is already determined at the zygote stage of development by its ______ chromosomes.

- -

gametes
diploid
sex

15. In the following cases, indicate the sex that will result:
(a) X-bearing sperm + X-bearing egg __________
(b) Y-bearing sperm + X-bearing egg __________

- -

female
male

16. Are there any other possible combinations of eggs and sperm? ________

- -

No, unless the meiosis is abnormal. For example, consider the rare case when both X chromosomes go to the same egg. This phenomenon, non-disjunction, explains how an XXY individual arises. (Cf. frames 7 and 12.)

17. Because of the nature of the sex chromosomes, the gametes from one sex actually defines the sex of the offspring. Which sex is that? ____________
Why? ____
____

- -

male
because the egg cell carries only X chromosomes but the sperm carries either X or Y (and the presence or absence of the Y chromosome determines sex).

18. If a sperm fertilizing an egg carries the X chromosome, the offspring will be a ________________ whereas if the sperm bears a Y chromosome, the offspring will be ______________ .

- -

female
male

19. The probability that a sperm will carry an X chromosome is __________. Therefore, the probability of a female offspring is ____________. By the same logic, the likelihood of a male child is __________. So as we discussed earlier in the chapter, the ratio of male to female offspring is approximately __________.

- -

0.5 (1 in 2 or 50%)
0.5
0.5
1:1

20. In some organisms, for example, birds and fishes, there is essentially a reversal of sex chromosomes so that the male has two identical sex chromosomes and the female has # ________________ sex chromosomes. In such cases, the sex of the progeny is determined by the ____________ ____________. Inasmuch as this book is devoted to the genetics of humans, this exceptional type of sex determination will not be considered further.

- -

two different
female or egg cell (cf. question 17)

21. Some applications of the binomial expansion to genetics were developed in Chapter 4 (#20–24). The same equation may be used to predict the most probable distribution of sexes in a family of any size. If we let X = males, Y = females, and let the exponent denote the number of progeny, then the expanded equation gives the sex distribution. For a two-child family, we expand: $(X + Y)^2$ to get $1X^2 + 2XY + 1Y^2$. This states that of four possibilities (i.e., 1 + 2 + 1), there is a one in four chance of having two boys (X^2), two chances in four of having one girl and one boy (XY), and # __________ chance(s) in # ______ of having two girls.

---

one
four

22. Work out the probable distribution of sexes for a five child family.

Expand $(X + Y)^5$ in either of the two following ways:

(A) $(X + Y)^5 = X^2 + 2XY + Y^2$

$$\frac{X + Y}{X^3 + 2X^2Y + XY^2 + X^2Y + 2XY^2 + Y^3 = 1X^3 + 3X^2Y + 3XY^2 + 1Y^3}$$

$$X^3 + 3X^2Y + 3XY^2 + Y^3$$

$$\frac{X + Y}{X^4 + 3X^3Y + 3X^2Y^2 + XY^3}$$

$$\frac{X^3Y + 3X^2Y^2 + 3XY^3 + Y^4}{X^4 + 4X^3Y + 6X^2Y^2 + 4XY^3 + Y^4}$$

$$\frac{X + Y}{X^5 + 4X^4Y + 6X^3Y^2 + 4X^2Y^3 + XY^4}$$

$$\frac{X^4Y + 4X^3Y^2 + 6X^2Y^3 + 4XY^4 + Y^5}{1X^5 + 5X^4Y + 10X^3Y^2 + 10X^2Y^3 + 5XY^4 + 1Y^5}$$

(B):

(1) First set up the terms of the expansion and recognize that the first and last coefficient must be 1. So, $1X^5 + X^4Y + X^3Y^2 + X^2Y^3 + XY^4 + 1Y^5$.

(2) Obtain the second coefficient from the first term as follows: multiply the coefficient by the power and divide by the term position so $(1 \times 5)/1$ gives 5, the coefficient of the second term.

(3) Obtain the third coefficient from the second term as above (i.e., coefficient $\times$ power $\div$ term or $5 \times 4 \div 2$ to get 10.

(4) Obtain the fourth coefficient from the third term (i.e., $10 \times 3 \div 3$ to get 10) and so on, to obtain $1X^5 + 5X^4Y + 10X^3Y^2 + 10X^2Y^3 + 5XY^4 + 1Y^5$.

Each term now corresponds to a possible combination of sexes in the family so, for example, Y^5 corresponds to five females (no males) and X^4Y corresponds to one female and four males. The total possible chances are the sum of the coefficients (i.e., $1 + 5 + 10 + 10 + 5 + 1$ or 32).

The chance of five girls ($1Y^5$) is 1 in 32; four girls and one boy ($5XY^4$ is 5 in 32; three girls and two boys ($10X^2Y^3$) is 10 in 32; two girls and three boys ($10X^3Y^2$) is again 10 in 32; one girl and four boys ($5X^4Y$) is 5 in 32; five boys ($1X^5$) is 1 in 32.

23. What is the likelihood of the particular combination of sexes in your own family?

Use a binomial expansion $(X + Y)^n$ in which n = the number of your siblings plus yourself.

NOTES

[1] We will use this generalization even though it is an approximation and should be qualified. Actually, the proportion of male and female infants is not exactly equal but typically slightly favors the males.

CHAPTER 7

X-Linkage

Not only do the sex chromosomes determine sex but they also lead to unique and interesting patterns of inheritance. Since the X and Y chromosomes are not homologous, they contain different genes. In fact, the Y chromosome appears to be essentially devoid of genes. Therefore, any male with a recessive gene carried on the X chromosome will show that recessive trait phenotypically. Genes on the X chromosome are said to be X-linked.

1. It was already seen (Chapter 6) that the X and Y chromosomes determine the _________ of the individual. Also, these chromosomes lead to unique modes of inheritance. Both of these phenomena are related to the fact that the X and Y chromosome are not _______________ and the Y chromosome apparently contains very few ___________ .

- -

sex
homologous
genes

2. Hemophilia, or "bleeder's disease," is inherited via a recessive gene on the X chromosome. A female may have as many as # ______ gene(s) for hemophilia while a male can have #_____ gene(s) for this trait. Thus, a female can be either homozygous or heterozygous with respect to the gene for hemophilia. How about a male? ______________________________
____

- -

two
one
A male can be neither homozygous nor heterozygous for an X-linked trait since he has only one X chromosome.

3. Since a male cannot be homozygous or heterozygous with respect to X-linked genes, he is sometimes described as "hemizygous." Let h signify the allele for hemophilia and H represent its dominant counterpart for normal (no hemophilia). List all possible genotypes for normal and hemophilic individuals of both sexes. ____

---

females: HH, Hh, hh
males: H, h (Other conventions will be discussed shortly.)

4. Several conventions are used for discussing X-linked genes. For example, a female homozygous for the normal allele may be shown as X^HX^H, which specifies that each X chromosome has the dominant ____________. What would be the comparable genotypes of a heterozygous and homozygous recessive female? ________________ , ________________.

---

allele
X^HX^h, X^hX^h

5. The phenotypes corresponding to the genotypes X^HX^H, X^HX^h, and X^hX^h are ____________________ , ____________________ , and ____________________ respectively.

---

normal
normal
hemophilic

6. In depicting the genotype of males using this convention, we must keep in mind that he has only # _____ X chromosome(s). A normal male may be shown as X^HY to specify that he has an X chromosome carrying the dominant allele, H, and a Y chromosome devoid of either form of this gene. Two other representations of X^HY are frequently used. They are $X^H\rightharpoondown$, and X^HO, in which the Y chromosome is shown as "$\rightharpoondown$" and "O." Occasionally,

X^HY is shown as $\frac{H}{\rightharpoondown}$, in which H signifies the X chromosome with the allele H and $\rightharpoondown$ signifies the Y chromosome. How would you designate the genotype of a hemophilic male?
Are there any other possible genotypes for hemophilia in a male?

---

one
X^hY or $X^h\rightharpoondown$ or X^hO or $\frac{h}{\rightharpoondown}$
no

7. The phenotypes represented by X^hY and X^HY are ______________ and ______________, respectively.

---

hemophilic
normal

8. A female gets #_____ X chromosome(s) from each parent. A male gets #_____ X chromosome(s) from his ______________.
Explain. ______________________________
____
____

---

one
one
mother
The female gets one X chromosome from each parent's gametes in the same way that she gets her autosomes. On the other hand, a male gets one Y chromosome from his father and one X chromosome from his mother.

9. Therefore, a hemophilic male inherits his abnormal gene from his ______________. A female with this disease must have received the allele, h, from # _____ parent(s).

---

mother
two (i.e., each)

10. Under what conditions could a hemophilic woman transmit the disease to her daughter? ______________________________

____

Only if her husband is also a hemophiliac.

11. Such a hemophilic daughter will be __________ genotypically.

X^hX^h

12. Can a hemophilic mother transmit the disease to her sons if she has a "normal" husband? ______________________________

Yes, since the son receives his one X chromosome from his mother.

13. Under what circumstances can a father transmit hemophilia to his sons?

____

To his daughters? ______________________________

____

____

He can't pass the disease to his sons. He can pass the disease to his daughters only if his wife contributes the recessive gene, h.

The pedigree chart below describes an X-linked recessive form of red-green color blindness (c). Questions 14-30 are based upon this chart. Shaded symbols are color blind.

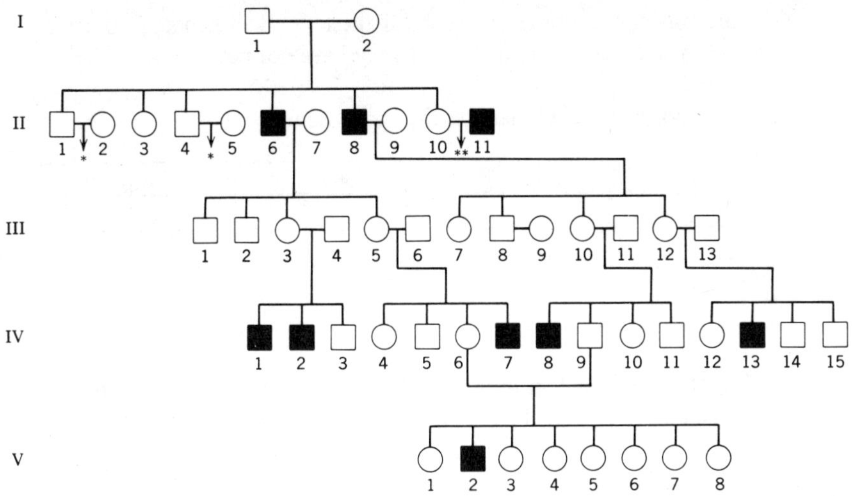

See note number 1 on page 79.

14. There are several subtle and unique features of this pedigree. First, only ________ have the trait.

---

males

15. A second intriguing observation is that sons of color-blind men are phenotypically (normal, color blind) .

---

normal

16. A third characteristic of recessive X-linked inheritance is that the trait frequently ________________ generations.

---

skips (or similar word)

17. Is it possible to deduce the genotypes of the males? ________________
Why? ____
____

Yes, since males carry only one gene for the trait, their genotype must be X^cY if they are color blind or X^CY if they have normal vision.

18. List the genotypes for all males in the above pedigree.

____

____

____

____

I-1, II-1, II-4, III-1, and all unshaded squares are not color blind and therefore, must be X^CY. II-6, II-8, II-11, and all shaded squares must be X^cY.

19. What generalization can be made about the genotypes of the females?

____

Since the females are all normal they must be either X^CX^C or X^CX^c.

20. Given that this form of color blindness is not common, what is the likelihood that females like II-5, II-7, and II-9 are carriers of the gene for color blindness? ______________________________
Explain. ______________________________

____

____

---

not likely
The trait is rare so there are relatively few such recessive alleles in the general population. Moreover, in each of these cases, there are no sons with the trait.

21. Which females in the chart are carriers? ______________________________
______________________________. Explain. ______________________________

____

I-2, III-3, III-5, III-10, III-12, IV-6
The mothers of all affected males must carry the recessive allele.

22. Which of the females would you expect to be genotypically X^CX^C?

_______________ Explain. _______________

II-2, II-5, II-10

Each of these women have numerous children but no affected sons.

23. In which cases is it impossible to deduce the genotype of females?

____

Why? _______________

II-3, II-7, II-9, III-7, III-9, IV-4, IV-10, IV-12, V-1, V-3, V-4, V-5, V-6, V-7, V-8. (See, however, question 20.)

There are too few male offspring to permit a judgment.

24. If #II-3 were known to be a carrier and she married a normal male, what would be the likely distribution of offspring?

____

50% of daughters: X^CX^C (normal)
50% of daughters: X^CX^c (normal)
50% of sons: X^cY (color blind)
50% of sons: X^CY (normal)

♀ \ ♂	X^C	Y
X^C	X^CX^C	X^CY
X^c	X^CX^c	X^cY

25. From the standpoint of a statistician, there is no case in the chart of an individual family of sufficient size to test this ratio. However, we can tabulate all the offspring of known heterozygous females and normal males. Which individuals matings may be tallied? _______________

Matings of parents: I-1 x I-2, III-3 x III-4, III-5 x III-6, III-10 x III-11, III-12 x III-13, and IV-6 x IV-9.

26. Are the data consistent with the above assignment of genotypes (frame 25)? ______________________________

Yes. Total offspring from this group: 29.
Predicted: 7 normal and 7 color-blind males, 7 normal and 7 heterozygous females.
Found: 8 normal and 8 color-blind males, 1 normal female, 1 carrier female, and 11 undeterminable females. The agreement with prediction is very good for the males, although inconclusive for the females.

27. If V-2 were to marry a carrier and have a large family, what would you predict about the ratio of genotypes? ______________________________

X^CX^c:X^cX^c:X^CY:X^cY
1:1:1:1

♀ \ ♂	X^c	Y
X^C	X^CX^c	X^CY
X^c	X^cX^c	X^cY

28. Even though we were told that the color-blind trait was X-linked, we could have inferred this from examination of the data. What are the essential arguments? ______________________________

---

1. The peculiar pattern of inheritance typifies an X-linked recessive gene cf. questions #14-16.
2. Color blindness is not autosomal dominant since neither parent of affected offspring show the trait.
3. Color blindness is not autosomal recessive since I-1, I-2, II-7, II-9, III-4, III-6, III-11, III-13, and others who marry into the family would all have to be carriers and this is unlikely.
4. If the trait were autosomal, it should affect females and males in equal numbers. (Still other arguments could be added.)

29. In X-linked traits, there should be a difference between reciprocal matings. That is, the ratio of phenotypes resulting from a mating of an affected male-X normal female should differ from a mating of affected female-X normal male. Is this seen in the pedigree above?________ Explain.________

____

____

- -

No, because there are no affected females but this is a useful test, especially if we are dealing with X-linked dominant traits.

30. Thus far, we have been considering X-chromosomal genes that are recessive. On the other hand, an X-linked trait may be controlled by a dominant allele. In the case of X-linked recessive genes, females were *not* affected. Would this be true with an X-linked dominant trait?________________

____

- -

No, the dominant gene will manifest itself in females as well as males. In fact, we will soon see that females are actually more likely to inherit an X-linked dominant trait than males. The pedigree for an X-linked dominant trait will be very different than one for an X-linked recessive trait.

The pedigree below shows the inheritance of hereditary enamel hypoplasia, E, in a hypothetical family. The symbol "*" is used as in the previous pedigree chart. Enamel hypoplasia, meaning that the enamel surrounding the teeth is very thin, is a relatively rare trait in the general population.

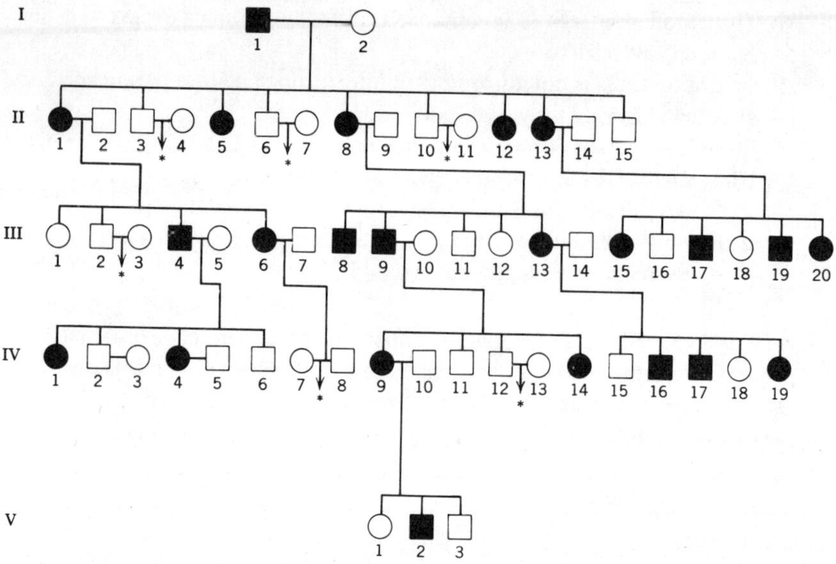

See note number 2 on page 79.

31. Considering the chart above, can the enamel trait be autosomal recessive? ________ Explain. ____________________

Not at all likely.
This hypothesis would require that #I-2, II-2, II-9, II-14, III-5, III-10, etc, were all carriers. This would be highly unlikely since this is an uncommon disease.

32. Could the trait be X-linked and recessive? ____________________

Not likely for the same reason as above. Also, if a female has an X-linked recessive trait, all of her sons will show this trait. Note offspring of #II-1, II-8, II-13, III-13, and IV-9. Moreover, with a rare X-linked recessive disease, we would not expect females to be affected.

33. Could the trait be autosomal and dominant? ____________________

At first glance this seems plausible, but read on.

34. If the trait were autosomal and dominant, then we might consider the total progeny from all matings of affected males or females with normal mates. Assume all affected individuals are heterozygous. Is this a reasonable assumption? ____________________

- -

Yes, they cannot be homozygous dominant since a homozygous person with this trait would have two affected parents. Moreover, if an individual is homozygous for a dominant gene, then all of his progeny will manifest this trait.

34A. What would you predict about the overall ratio of affected : normal progeny? ____________

- -

1:1

34B. What do you find and conclude? ____________________

____

- -

22 with the trait, 19 normal. This is good agreement with the predicted ratio. So far, all the observations are consistent with the autosomal dominant hypothesis.

35. If the trait is autosomal dominant, there should be a ____ ratio of affected males to affected females. The actual numbers are #____ affected males and #____ affected females. What does this suggest? ____________

____

- -

1:1
9
14
Since this ratio is closer to 1:2 than 1:1, the hypothesis of autosomal dominance may be questionable.

36. Considering all matings of the type ■—○ , what would you predict for phenotypes of offspring based on the autosomal-dominant hypothesis? ________ Explain how you got this ratio. ________________________________

____

1:1
The male parents are Aa and the females are aa. From a Punnett square, we get the following genotypes:

♀ \ ♂	A	a
a	Aa	aa

36A. Does the pedigree chart support or contradict your prediction? ________

____

---

contradicts (From matings of affected males and nonaffected females, we find nine females with the enamel trait and no males with enamel hypoplasia. This is a far cry from the 1:1 ratio predicted.)

36B. What can we conclude about the autosomal-dominant hypothesis?

____

The autosomal-dominant hypothesis appears to be incorrect as the probability of having nine daughters (all affected) and eight sons (all normal) is very small. We can actually calculate how small this likelihood is by the method that follows.

37. If we know the probability of occurrence for each of a series of events, we can easily determine the likelihood of all of the events occurring. That chance is equal to the product of all the probabilities. In tossing coins, what is the chance of tossing a head?____________ What is the chance of tossing three heads in a row? ________________

---

1 in 2
1 in 8 (i.e., $1/2 \times 1/2 \times 1/2$)

38. Now let us apply this "product probability law" to the autosomal-dominant hypothesis. Previously, we predicted that offspring from matings of the type ■——○ should be phenotypically 1:1 since we were dealing with a test cross (Ee × ee). Would these results be dependent on sex?

____

no (since we are testing an autosomal hypothesis)

38A. The actual results from the pedigree chart were nine daughters affected, no daughters with normal enamel, eight sons with normal enamel, and no sons with enamel hypoplasia. If the offspring had included only one female, what is the likelihood that she would have had the enamel trait? ________ Suppose the offspring included nine daughters, all affected, how likely would that be? _______________

1 in 2 (or 50%)
(1/2) (1/2) (1/2) (1/2) (1/2) (1/2) (1/2) (1/2) (1/2) or $(1/2)^9$ or one chance in 512

38B. Now, consider the actual cases of nine daughters (affected) and eight sons (normal) and no other phenotypes: what is the probability of all these events? _______________

The probability of having nine affected daughters and eight normal sons would be $(1/2)^{17}$ or only one chance in 131,072! The autosomal dominant hypothesis is clearly untenable.

39. Consider the possibility that hereditary enamel hypoplasia is an X-linked dominant trait. Are there any data in the pedigree that contradict this hypothesis? ________ Does this prove that the trait is X-linked dominant?

____

no
no, not by itself

40. In matings of ●—□, the sexes of the offspring should have enamel hypoplasia in a____:___ratio.

- -

1:1

41. In the reciprocal matings (■—○), what would you predict based on the X-linked dominant hypothesis?______________________________

____

- -

All daughters and no sons will have enamel hypoplasia as shown by the Punnett square.

♀ \ ♂	X^E	Y
X^e	$X^E X^e$	$X^e Y$

42. Since both reciprocals are found in the pedigree, what can we predict about the proportions of females and males with the trait (if an X-linked dominant gene causes enamel hypoplasia)?______________________________

____

- -

There should be more females with the trait than males.

42A. Is this prediction supported by the pedigree chart?

____

- -

Yes. The chart shows 9 males and 14 females with hereditary enamel hypoplasia.

43. If a trait is X-linked dominant, it <u>(can, cannot)</u> be transmitted from father to son. Careful analysis of the chart for such a pattern shows

____.

- -

cannot
it never occurs

44. In summary, the disease under study appears to be inherited via a
(X-linked, autosomal) ________ (dominant, recessive) ________ gene.

X-linked
dominant

44A. List the arguments that support this conclusion.

---

1. The trait cannot be autosomal recessive (frame 31).
2. The trait cannot be X-linked recessive (frame 32).
3. Although some aspects of the pedigree can be explained by a hypothesis of autosomal dominant (frames 33, 34), further analysis refutes the autosomal dominant hypothesis (frames 35, 36, 38).
4. There is nothing in the chart that would refute the X-linked dominant hypothesis (frame 39).
5. More females than males have the trait (frames 40-42).
6. The trait is never passed from father to son (frame 43).
7. All daughters of affected males have the trait.

BRIEF REVIEW OF CHAPTER 7

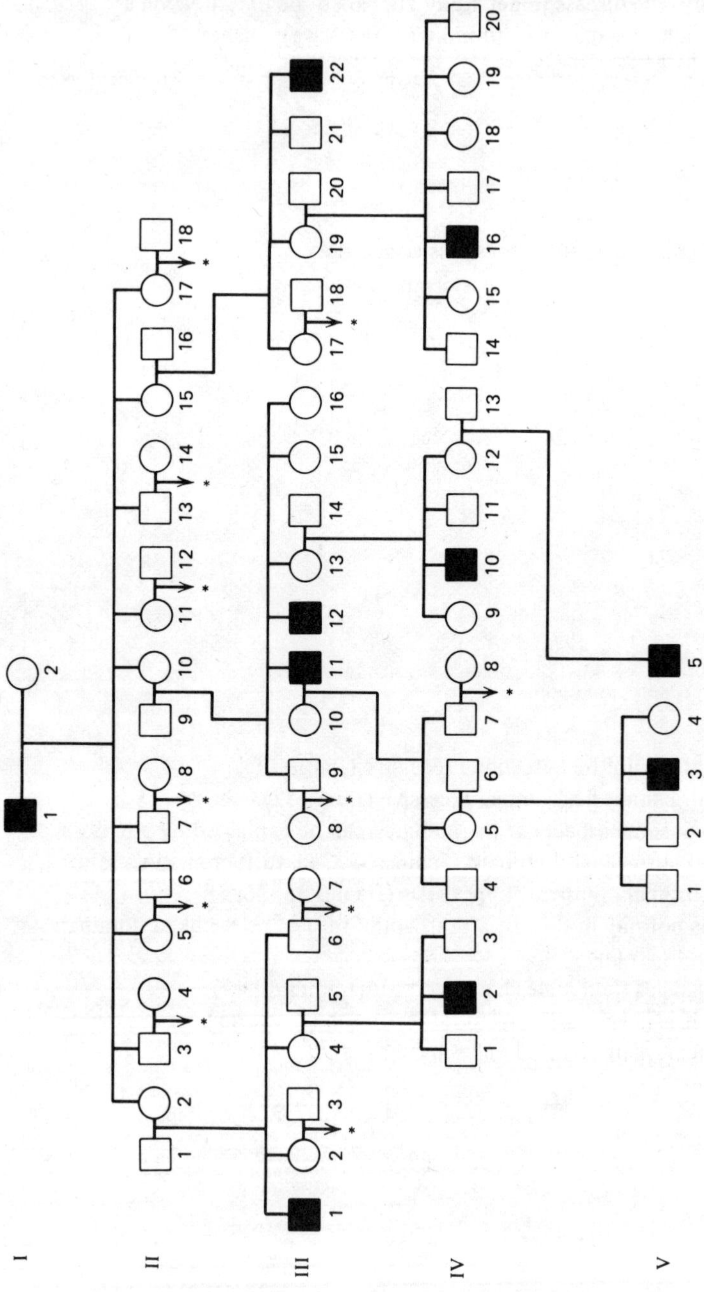

The asterisk (*) symbol in the chart above is used as in the previous pedigree.

45. Study the above pedigree chart. What is the most probable mode of inheritance? ______________________

Provide convincing arguments in support of your conclusion. (We are dealing with a rare trait.)

---

1. Eliminate autosomal or X-linked dominant hypothesis since parents of offspring with the trait do not have the trait. Moreover, these hypotheses would not explain why only males manifest the trait.
2. Eliminate the autosomal recessive hypothesis because (a) the trait does not occur in females, and (b) it would be unlikely for numerous individuals such as #I-2, II-1, II-9, and II-16 to be carriers of a rare trait.
3. Note the typical pattern of an X-linked recessive trait (i.e., only males have the trait; the trait frequently skips generations).
4. From a total of eight crosses of presumed carriers with normal males, there were 20 sons. If the X-linked recessive hypothesis is valid, then we may predict that half of the sons would be affected with the trait. In the chart, there are 11 normal sons and 9 affected sons. These numbers are in very good agreement with the prediction.
5. Based on all of the arguments, we may conclude that the trait is almost certainly X-linked recessive.

The following pedigree also describes a hypothetical trait which may be presumed to be very rare.

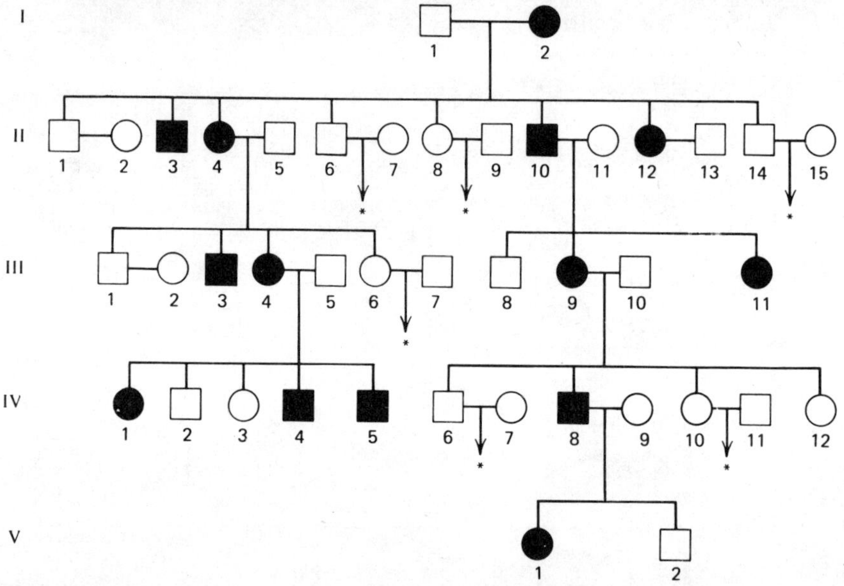

*The asterisk symbol is used as in previous charts.

46. What is the most probable mode of inheritance of the trait in the pedigree chart? Explain. ____________________

1. Eliminate the autosomal recessive hypothesis because it would imply numerous carriers of what is known to be a rare condition (cf. previous question).
2. Eliminate the X-linked recessive possibility because (a) it occurs in females frequently and because (b) women with normal fathers have the trait.
3. An autosomal dominant mode seems to be a possibility. If we assume that all crosses of affected individuals with normal mates correspond to heterozygous affected and homozygous recessive normal individuals, then we may predict that (a) half of the offspring should be normal and the rest should be affected and (b) that there should be as many affected males as females. This is seen to be true. In the six matings of affected and normal individuals, there are 26 offspring of which 13 are normal and 13 are not. The affected offspring include males and females in approximately a 1:1 ratio. The data are not extensive but they are consistent with an autosomal dominant pattern.
4. We should also consider the possibility that the trait could be X-linked and dominant. There is, surprisingly, no way to rule out this possibility with the available information. The most useful kind of test for an X-linked dominant trait is the cross of an affected male with a normal female; such crosses should lead to only two classes of progeny: normal males and affected females in a 1:1 ratio. In the pedigree chart, we find only two such crosses with five offspring but the offspring include two normal males and three affected females. These results are consistent with an X-linked dominant hypothesis.
5. In summary, we note that two distinct modes of inheritance are consistent with the data. We might be inclined to lean very slightly toward the X-linked dominant hypothesis because of the results of crosses of affected males with normal mates. However, the data are insufficient to permit a definitive proof of either the autosomal or X-linked model.
6. This problem was chosen deliberately because of its ambiguity. It is fairly typical of many human case histories where the family size is small. The solution to such a pedigree problem is sometimes made

more complicated by additional variables such as dubious and unverifiable information or illegitimacy in a family. The puzzled investigator pursues his analysis with a cold, critical logic and makes use of the literature on similar cases for guidance.

NOTES

[1] The asterisk signifies that the offspring of these individuals are (or were) numerous and are (or were) not color blind. (It should be apparent that in a pedigree with five generations, many of the individuals shown are deceased).
[2] This couple had five normal sons (no daughters) and these sons fathered numerous normal children.

CHAPTER 8

Autosomal Linkage

In Chapter 4, we examined genetic matings involving two pairs of alleles located on different chromosomes. The observed 9:3:3:1 ratio from a dihybrid cross results from the random segregation of homologous autosomal chromosomes during meiosis. We point out that the total number of human genes has been estimated to be of the order of 10^5. Since there are only 23 pairs of human chromosomes, each chromosome must bear many genes. When two or more genes are carried by the same chromosome, the pattern of inheritance may be affected. Such genes are said to be linked because they tend to segregate together in meiosis. The inheritance of linked genes is the subject of this chapter.

1. In a mating of the type AaBb x AaBb, we may normally predict a ratio of phenotypes in the offspring of ________ . This kind of mating is called a ______________ cross. The results depend on the assumption that #______ pairs of chromosomes are involved.

- -

9:3:3:1 (Cf. Chapter 4)
dihybrid
two

2. In the above case, each diploid cell of the parents must have #________ chromosomes carrying the relevant genes. The gametes produced by these parents each carry #________ of the genes. The possible combinations of genotypes in a sperm or egg are ______________________________

- -

four
two

AB, Ab, aB, ab, but not AA, Aa, BB, Bb because of segregation of homologs. (If nondisjunction occurs as alluded to in Chapter 6, frame 16, a gamete could be genotypically Aa or BB. Such an event is unlikely and will not be considered further in this text.)

3. In the problem being considered, the alleles A and B are assumed to be on different or nonhomologous chromosomes. In other words, alleles A and B are not ____________ on the same chromosome.

---

linked

4. It is, however, possible for alleles A and B to be linked to a particular ____________ since there are numerous ____________ on each chromosome.

---

chromosome
genes (or alleles)

5. Let us assume that genes A and B are linked on one chromosome (for example, chromosome 20) and that the linkage cannot be broken. If the individual with this chromosome is known to be dihybrid, what genes must the homologous chromosome contain? ____________

---

a and b (since that individual must have these recessive alleles to be AaBb)

5A. What kinds of gametes could this person produce, ____________

---

either AB or ab (in other words, a gamete would contain a chromosome with the genes A and B on it or a chromosome with a and b)

6. Pursuing the autosomal linkage assumption, the person of genotype AaBb carries these alleles on two homologous chromosomes. To designate such linkage, the symbol "<u>AB</u>" denotes a chromosome (i.e., the horizontal line)

with linked genes A and B on it. An individual homozygous for the dominant genes would be shown as "$\frac{AB}{AB}$". How would you represent a homozygous recessive individual with linked genes? ____________
Give two possible combinations for a dihybrid individual? ____________

____

- -

$\frac{ab}{ab}$

$\frac{ab}{AB}$ or $\frac{Ab}{aB}$ (Note: $\frac{ab}{AB}$ is equivalent to $\frac{AB}{ab}$.) A heterozygous individual can be either $\frac{AB}{ab}$ or $\frac{Ab}{aB}$ but not both.

7. A man who is genotypically $\frac{AB}{ab}$ produces sperm of what genotype(s)?

____

- -

$\underline{AB}$: $\underline{ab}$ / 1:1

8. Suppose this man marries a woman of the same genotype, predict the genotypic ratio of offspring. ____________

____

- -

$\frac{AB}{AB}:\frac{AB}{ab}:\frac{ab}{ab}$ / 1:2:1

♀ \ ♂	$\underline{AB}$	$\underline{ab}$
$\underline{AB}$	$\underline{AB}$/ /$\underline{AB}$	$\underline{AB}$/ /$\underline{ab}$
$\underline{ab}$	$\underline{AB}$/ /$\underline{ab}$	$\underline{ab}$/ /$\underline{ab}$

9. Contrast the ratio of offspring phenotypes for a dihybrid mating with linked genes with (1) the ratio for a dihybrid mating with nonlinked or independent genes and (2) the ratio for a monohybrid mating.

____

____

____

____

- -

The ratio for the dihybrid with linked genes differs markedly from the usual dihybrid ratio (independent genes) of 9:3:3:1 but is identical to the monohybrid ratio. This is because linked alleles segregate together in meiosis.

Assume that the following pedigree chart demonstrates the inheritance of the trait polydactyly (i.e., the presence of extra fingers and/or toes).

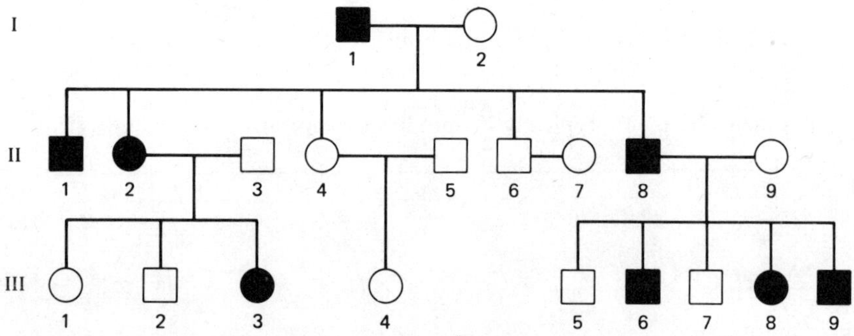

10. What is the probable mode of inheritance of polydactyly?

____

---

autosomal dominant (to review the arguments, see, for example, Chapter 7, frames 45 and 46.)

11. Assume now that the dominant allele for carrying the Rh factor (i.e., determining Rh+) is linked to the allele for polydactyly and that individual #I-1 is Rh+. What other individuals would be Rh+? _______________

____

(Assume that #I-2, II-3, II-5, and II-9 are Rh–.)

#II-1, II-2, II-8, III-3, III-6, III-8, III-9. Note: up until now, we have been assuming that linked genes are always transmitted together. The assumption is not necessarily valid.

12. Suppose we are told that #II-2 has six fingers and is Rh–, and that #III-6 has the normal number of fingers and toes and is Rh+. Is this consistent with our prediction for linked genes? ______
How might these results be explained? ______

______

______

(There are actually several possible explanations and these have not been considered yet in detail. Take some guesses.)

---

no
First, the linkage may not be "firm," that is, the chromosome may be broken. This is an important possibility and will be analyzed further shortly. Secondly, a gene may actually be present and not manifested. This is called "lack of penetrance." Thirdly, and possibly without the knowledge of the genetic analyst, individuals #II-2 and III-6 may be adopted or illegitimate. Another possibility is that there was an error in reporting a phenotype; human fallibility must always be kept in mind.

13. In meiosis, two homologous chromosomes may sometimes "cross over" (i.e., break and exchange segments containing several or many alleles). For example, what genotype would result from a rupture and exchange of chromosomes between A and B in $\frac{AB}{ab}$? ______

$\frac{Ab}{aB}$

14. If there is no crossing over between A and B, the number of possible kinds of gametes is #____; they are____and____ genotypically. Assuming that crossing over may occur, there should be #____kinds of gametes, that is,

______

---

two
AB
ab
four
AB, ab, Ab, aB

15. Consider two homologs with the following configuration: $\frac{RsT}{rSt}$.
What kinds of variants ("recombinants") may result from crossing over?

____

$\frac{Rst}{rST}$, $\frac{rsT}{RSt}$, and $\frac{RST}{rst}$

16. How could the recombinant $\frac{RST}{rst}$ have formed? ____________________

____

____

This combination would result from breakage and recombination of homologous chromosomes at two points (i.e., double crossing over). This is sometimes shown schematically as follows:

R s T   R S T
XX   ⟶
r S t   r s t

17. What kinds of gametes (i.e., genotypes) may result if crossing over may occur in this instance? (from $\frac{RsT}{rSt}$)

____

For two pairs of linked alleles, we noted that four gamete genotypes could occur with crossing over. (Cf. questions 13 and 14.) For three pairs of alleles, we found #_____ kinds of gamete genotypes. (Cf. question 16.) For n pairs of alleles, we may predict #_____ gamete genotypes.

RsT, rSt, RST, rst, Rst, rST, RSt, rsT
eight
2^n

PAIRS OF ALLELES	GAMETE GENOTYPES
2	4 or 2^2
3	8 or 2^3
n	2^n

18. Darwin's theory of evolution presupposes variation of phenotypes within a species. Such variation may stem from changes of specific genes (i.e., "mutations"), and this will be discussed in Chapter 11. Can you suggest another source of variation? ____________________

---

crossing over (recombination)

19. We have pointed out already that for two homologs with n pairs of alleles, crossing over may provide 2^n gamete genotypes. What is the implication for evolution of the fact that a typical human chromosome probably has over a thousand genes?

____

____

____

____

The potential for variation of genotypes and, consequently, phenotypes that may result from crossing over is almost infinitely large. Recombination provides an important argument in support of Darwinian evolution.

20. Assume that crossing over may occur at any point along the chromosomes. Therefore, it is logical that two alleles at opposite ends of a chromosome are more likely to be recombined than two adjoining alleles since, in the former cases (two genes far apart), there are many points along the length of the chromosomes at which they could break and exchange fragments, but in the latter case (two genes close together) there are few points at which recombination could occur. In general, results of matings involving linked alleles are consistent with this expectation even though there are regions of the chromosome in which crossing over is inhibited. If we find a low proportion of recombinants resulting from matings that involve two linked genes, what may we conclude about these two genes?

____

---

they must be close to each other

21. In this way, geneticists can measure the approximate distance between linked genes. This is called "mapping." Why is it that such distances are only approximations?

____

____

____

If two linked genes are in an area of the chromosome in which crossing over is inhibited, the number of recombinants will be low and the apparent map distance will be less than the true distance.

22. We know that the genes for hemophilia, h, red-green color blindness, c, and a kind of congenital deafness, d, are all linked on the X chromosome. If the frequency of crossing over is approximately 8% between alleles h and c, and 17% for alleles h and d, what can we infer? ______________

____

____

Alleles h and d are about twice as widely separated as alleles h and c. Possible configurations are d _______ h ___ c or h ___ c ___ d.

23. In man, most of the chromosome mapping has been done on the X chromosome. Why? (Guess.) ______________________

____

____

It is easy to infer that specific genes are on the X chromosome because traits on the sex chromosomes show unique patterns of inheritance. The assignment of genes to specific chromosomes is usually a prerequisite to any mapping study. With the exception of the X chromosome, this has been very hard to achieve in humans. However, several new research techniques are showing promise for the future of human chromosome mapping.

BRIEF REVIEW OF CHAPTER 8

24. Assume that the traits widow's peak (p) and attached ear lobes (a) are carried by linked recessive alleles. What would be the genotypic and phenotypic results of the following mating if (1) the linkage is not broken, (2) crossing over occurs. $\frac{pa}{pa}$ x $\frac{Pa}{pA}$ ________________________

---

1. No crossing over (a) the gametes for the homozygous parent are pa and the gametes of the dihybrid parent are Pa and pA.
(b) The ratio of genotypes for the offspring would be $\frac{pa}{Pa}:\frac{pa}{pA}$ / 1:1.

♀ \ ♂	Pa	pA
pa	Pa/ /pa	pA/ /pa

(c) The phenotypic ratio of the offspring is 1:1 straight hairline and and attached ear lobe to widow's peak and "hanging" or "free" ear lobes.

2. Crossing over may occur. (a) The gametes from the parent that is $\frac{pa}{pa}$ are still all pa. It is possible for crossing over to occur between the two chromosomes but recombination would not lead to any change.
(b) The gametes from the other parent may be Pa or pA or PA or pa. The first two would be gamete genotypes if crossing over does *not* occur and the last two are recombinants.
(c) A Punnett square may be used giving the genotypic proportions:

♀ \ ♂	Pa	pA	PA	pa
pa	Pa/pa	pA/pa	PA/pa	pa/pa

(d) The possible genotypic classes based on the tabulated data above are attached ear lobes, widow's peak; hanging ear lobes, straight hairline; hanging ear lobes, widow's peak; and attached ear lobes, straight hairline. The precise ratios would depend on how closely linked the genes are and on their location on the chromosome.

CHAPTER 9

Quantitative or Polygenic Inheritance

While it is true that many traits are determined by a single pair of alleles, a number of genes may contribute to a single trait. These individual gene contributions to a trait may or may not be equal. Such inheritance is called "polygenic" or quantitative. When several genes contribute to a single trait, the result is often a measurable range or distribution of phenotypes. Examples of polygenic traits of humans include size, weight, and intensity of pigmentation. In some polygenic traits, such as IQ, the impact of environment is especially important. The exact mechanisms by which polygenic traits are controlled is not well understood but presumably is based on the fundamental biochemical nature of life. For instance, a trait such as height is dependent on growth, which requires hormonal action and the capacity to utilize nutrients (i.e., metabolism). Metabolism requires a variety of enzymes. Each enzyme is under genetic control, and each hormone is under the control of one or more genes so it becomes easy to visualize how a trait can be polygenic. The goal of this chapter will be to consider the implications of polygenic traits for patterns of inheritance and to contrast such inheritance with the simpler examples we have been considering thus far.

1. Assume that height in humans is determined by a single pair of alleles with T tallness dominant over t. If this were true, the offspring of a homozygous tall woman and a homozygous short man would be genotypically ______ and phenotypically ________ .

Tt
tall

2. In reality, the offspring of such matings tend to be intermediate in size. What is the simplest possible explanation for this observation?

____

---

incomplete dominance (However, read on.)

3. Based on your experience, what might you predict about the height of a child of homozygous tall and homozygous short parents if he were raised on a deficient diet? Explain.

____

____

____

In this case, even if the gene for tallness were dominant, an individual deprived of an adequate diet probably would not be tall. This is explained by the fact that environment influences genetic potential.

4. Most geneticists agree that IQ is determined partly by genes and partly by environment. A child of a couple of high IQ who receives a poor education is likely to have <u>(lower, higher)</u> IQ than his parents. Why?__________

____

---

lower
because of inadequate environment

5. Genes may be affected by nutrition, temperature, disease, and other external factors that constitute the ____________________. These factors help to explain how individuals with identical genotype can vary in ____________________. Such variation is called expressivity. It is also possible for an individual not to express a gene at all, and this is known as lack of penetrance. The general explanation of penetrance and expressivity is that they are due to either (1) environmental influences (including those that occur in the embryonic stages) or (2) modification of gene expression by other genes (i.e., modifying genes as discussed in Chapter 5) or a combination of environmental and genetic effects.

---

environment
phenotype

6. The condition of polydactyly (multiple fingers or toes) is controlled by a dominant gene. Polydactylic individuals will sometimes have a digit that is only partly formed exemplifying variable ____________________ . Theoretically, polydactyly should not skip generations because it is carried by a dominant gene. Occasionally, polydactyly will skip a generation and an individual may carry this dominant gene and yet have normal hands and feet. This would be due to a lack of ____________________ . The puzzling phenomena of expressivity and lack of penetrance may have an environmental basis.

---

expressivity
penetrance

7. Returning to the hypothetical mating above of a homozygous tall woman and homozygous short man, the two simplest expalnations for a medium-sized son are __________________ __________________ or complete dominance coupled with poor ____________________ .

---

incomplete dominance
environment

8. If we knew that the medium-sized son was raised under optimal conditions, then the fact that he is shorter than his mother would suggest __________________ __________________ as an explanation. This inference may be tested if this medium-sized son mates with a genotypically identical (medium-sized woman). Assuming that this occurred, give the genotypes and phenotypes for their offspring with predicted ratios.

____

____

---

incomplete dominance
TT, Tt, tt; 1:2:1
tall, short, medium; 1:1:2

9. Instead of falling into three discreet phenotypes, the progeny of medium-sized parents tend to range in size from somewhat smaller to somewhat larger than the parental height even when the external environment is constant. These results (confirm, refute) the hypothesis of incomplete dominance.

---

refute (Therefore we must look for another explanation.)

10. It is possible for more than one pair of alleles to determine height. If there are two pairs of alleles each contributing equally to height (T, t, H, h), then the progeny of two hybrids would consist of #_____ genotypes. (Assume constant environment.)

---

nine (The Punnett square is given below. You would obtain the same answer by expanding $[(2X) + (1Y) + (1Z)]^2$ as we did in Chapter 4, frame 24.)

	HT	Ht	hT	ht
HT	HHTT	HHTt	HhTT	HhTt
Ht	HHTt	HHtt	HhTt	Hhtt
hT	HhTT	HhTt	hhTT	hhTt
ht	HhTt	Hhtt	hhTt	hhtt

11. Assume the alleles, T and H, each contribute three units to height and the alleles, t and h, each contribute one unit to height. What phenotypic classes would result? ____________________ In what ratio? ____________

The phenotypic classes are: 12, 10, 8, 6, and 4 units in a ratio of 1:4:6:4:1.

GENOTYPE	FREQUENCY	PHENOTYPE
HHTT	1	12 units
HHTt	2	10
HhTT	2	10
HhTt	4	8
HHtt	1	8
hhTT	1	8
Hhtt	2	6
hhTt	2	6
hhtt	1	4

11A. What assumption is implicit about the environment?

The implicit assumption is that environment is not affecting the expression of the genotypes or that any environmental effect is the same for all genotypes.

11B. Can you suggest another assumption that is required?

We assume that there are no modifying genes.

12. Contrast the results of such a cross with the results from a monohybrid cross (assume incomplete dominance).

In the former we would get a greater range of values (i.e., a distribution of phenotypes); in the latter, we get three distinct phenotypes.

13. It has been noted that in polygenic inheritance, the phenotypes tend to be spread over a range. Such data are conveniently represented as a bar graph or histogram. For example, if your grades in a series of quizzes included 3 As, 7 Bs, 11 Cs, 5 Ds, and 2 Fs, a bar graph as shown below would provide a convenient picture of your standing in the course.

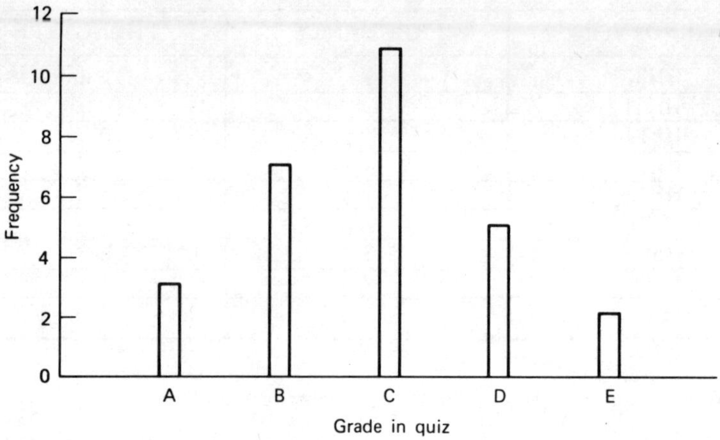

14. Provide a bar graph for the polygenic cross in frame 11.

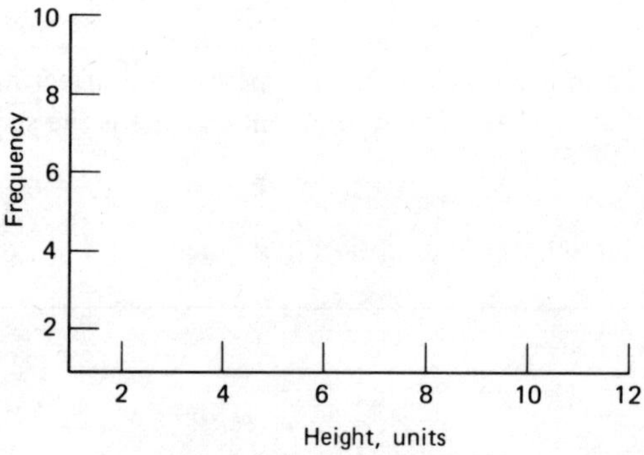

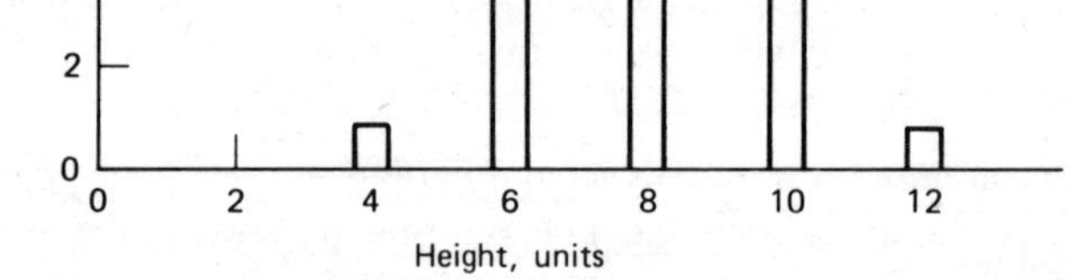

15. In the problems above, we considered the impact of either one or two genes on height. True polygenic or quantitative inheritance entails several genes, each contributing a small amount to a phenotype. In this chapter, we have considered the effect of matings of Aa x Aa and AaBb x AaBb. What would be the genotypic ratios for offspring resulting from a mating of AaBbCc x AaBbCc? ______________________________

____

____

____

A few hints may be helpful in solving this problem because a Punnett square would be awkward (64 little boxes to fill in). We can also obtain the genotypic ratios by expanding $[(2X) + (1Y) + (1Z)]^3$ as in frame 10 and grouping the 27 different terms. The term with the highest coefficient will give the trihybrid (e.g., AaBbCc) frequency. The terms with the next highest frequency will give the frequency of each dihybrid possibility (e.g., AABbCc, AaBBCc, AaBbcc). The third most frequent combinations will give the frequency of all monohybrids, and the least common terms will give the frequencies of the various possible homozygous types (e.g., AABBCC, AAbbcc, aaBBcc).

AABBCC	1	AAbbcc	1
AABBCc	2	AaBBCC	2
AABBcc	1	AaBBCc	4
AABbCC	2	AaBBcc	2
AABbCc	4	AaBbCC	4
AABbcc	2	AaBbCc	8
AAbbCC	1	AaBbcc	4
AAbbCc	2	AabbCC	2

AabbCc	4	aaBbCc	4
Aabbcc	2	aaBbcc	2
aaBBCC	1	aabbCC	1
aaBBCc	2	aabbCc	2
aaBBcc	1	aabbcc	1
aaBbCC	2		

16. A third method of analyzing the above trihybrid cross may be the most appealing. This approach takes advantage of two principles we have studied: (1) each of the gene pairs in a trihybrid cross behaves independently (Chapter 4) and (2) the likelihood that a series of events will occur is the product of the individual probabilities (Chapter 7, frames 37 and 38). So, for example, to determine the chance of the phenotype aaBbCc resulting from a trihybrid mating, we would have to surmise that the chance of getting aa (from Aa x Aa) is 1/4, and the chance of getting either Bb or Cc is 1/2 (from Bb x Bb or Cc x Cc). The probability of getting all three is 1/4 x 1/2 x 1/2 or 1/16. What is the probability of getting each of the following genotypes? 1. aabbcc ______
2. AaBbCc ______
3. AABBCc ______

---

1. The chance of getting aa (from Aa x Aa) is 1/4. Similarly, the chance of getting bb is 1/4 and of getting cc is 1/4. The probability of getting aabbcc, then, is $(1/4)^3$ or 1/64.
2. 1/8 (2/4 x 2/4 x 2/4)
3. 1/32 (1/4 x 1/4 x 2/4)

17. What would be the phenotypic ratios for height in problem 15 if we assume (1) incomplete dominance and (2) that A, B, and C each contribute two units to height and that a, b, and c each contribute 1/2 unit to height?

______

______

______

______

HEIGHT, UNITS	FREQUENCIES
12 units	1
10.5	6
9	15
7.5	20
6	15
4.5	6
3	1

18. Design a bar graph to show the frequencies of phenotypes in the previous frame.

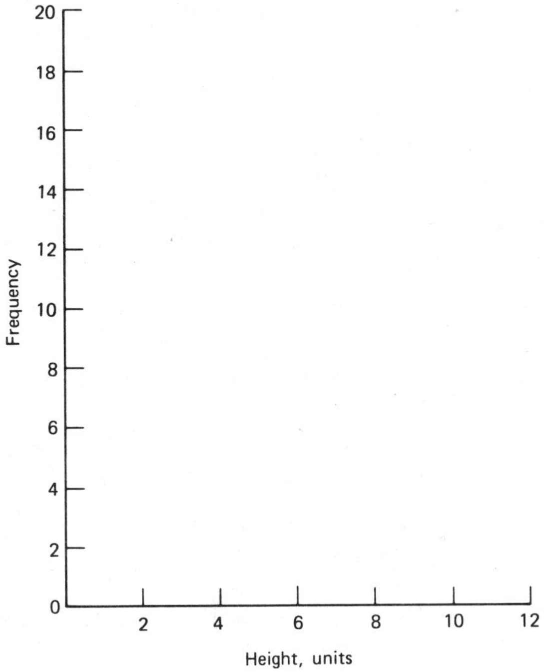

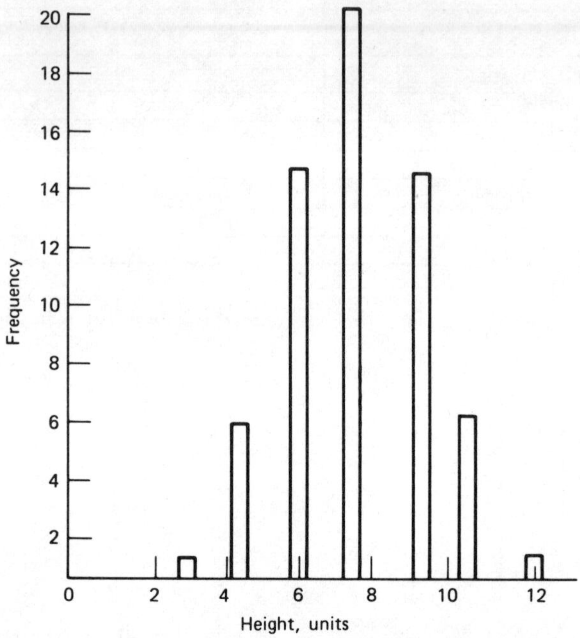

19. What additional assumptions were implicit in obtaining the phenotypic results above? ______________________________

We assumed (1) the absence of modifying genes and (2) that the environmental impact on phenotype was either nil or was equal for each genotype.

20. Compare the phenotypic ratios for the three kinds of matings treated in this chapter: Aa × Aa, AaBb × AaBb (problem 11), and AaBbCc × AaBbCc (problem 17). Can you detect the mathematical basis for the ratios obtained? (Clue: consider the binomial expansion.) ____________________

We are dealing with binomial expansions of $(X + Y)^n$ where n = the number of alleles that determine height. (Cf. Chapter 4).
(a) In the monohybrid cross, the phenotypic ratio is 1:2:1. These numbers could also be obtained by expanding $(X + Y)^2$ [n = 2 since there are two alleles determining height] $(X + Y)^2 = \underline{1}X^2 + \underline{2}XY + \underline{1}Y^2$.
(b) In the dihybrid cross, the phenotypic ratio is 1:4:6:4:1. We could get this from expanding $(X + Y)^4 = \underline{1}X^4 + \underline{4}X^3Y + \underline{6}X^2Y^2 + \underline{4}XY^3 + \underline{1}Y^4$.
(c) In the trihybrid cross, the phenotypic ratio is 1:6:15:20:15:6:1. We could get this as follows: $(X + Y)^6 = \underline{1}X^6 + \underline{6}X^5Y + \underline{15}X^4Y^2 + \underline{20}X^3Y^3 + \underline{15}X^2Y^4 + \underline{6}XY^5 + \underline{1}Y^6$.

21. As we increased the number of gene pairs in matings of heterozygotes, we found the following:

TYPE OF MATING	GENE PAIRS IN EACH HETEROZYGOUS PARENT	CLASSES OF GENOTYPES IN OFFSPRING
Monohybrid	1	3
Dihybrid	2	9
Trihybrid	3	27

Can you surmise the mathematical relationship between gene pairs in the heterozygous parents and the number of different kinds of genotypes in the offspring? ____________________

____

____

____

The number of kinds of genotypes in the offspring is 3^n where n is equal to the number of gene pairs in each heterozygous parent. (If you want to review this kind of reasoning, cf. Chapter 4, frames 20-32.)

22. If we increase the number of heterozygous gene pairs to 10, the number of different kinds of genotypes would be 3^{10} or a staggering 59,049. As we increase the number of genes involved in a polygenic trait such as height, what happens to the potential for phenotypic variability in the offspring?

____

Not only do the number of genotypic classes increase, but also more phenotypes are possible since the phenotypes are determined by the genotypes.

23. Therefore, the more classes of phenotypes we have, the more bars we have in bar graphs and the graphs become smoother, continuous, and less step-like. Something else happens to the bar graphs as we increase the number of genes in polygenic matings. Compare the bar graph in frame 14 (dihybrid parents) and frame 18 (trihybrid parents). What seems to be happening to the shape of the curve? ____________________

____

- -

The graphs become steeper as the number of heterozygous gene pairs increase.

24. Speculate on the nature of a bar graph for a polygenic trait where n, the number of pairs of alleles, is large.

____

____

- -

The curve will be very smooth and will appear like a normal distribution curve (i.e., a bell-shaped curve).

25. Contrast the phenotypic results of matings involving polygenic and nonpolygenic traits with respect to the distinguishability of the phenotypes. ______________________________

____

____

____

____

____

- -

This question may be answered in numerous ways, but the key points are the following. In polygenic inheritance, the individual phenotypes form a range; they are said to be "continuous." Polygenic phenotypes are usually amenable to measurements; they involve intensities or quantities and are, therefore, often called "quantitative traits." On the other hand, nonpolygenic traits give phenotypes qualitatively distinct from each other and are therefore "discontinuous."

26. We have been considering matings of monohybrids (Aa), dihybrids (AaBb), and trihybrids (AaBbCc). Indicate how each of these heterozygous individuals could have homozygous parents.

An Aa person could have parents who are AA and aa. Correspondingly, an AaBb individual's parents could be AABB and aabb (or AAbb and aaBB). Finally, an AaBbCc individual could result from AABBCC and aabbcc parents (among other possibilities).

26A. Fill in the following table.

GENOTYPES OF		PROBABILITY GENOTYPE OF CHILD IDENTICAL TO GRANDPARENT
BOTH PARENTS	A GRANDPARENT	
Aa	AA	
AaBb	AABB	
AaBbCc	AABBCC	

GENOTYPES OF		PROBABILITY GENOTYPE OF CHILD IDENTICAL TO GRANDPARENT
BOTH PARENTS	A GRANDPARENT	
Aa	AA	1/4
AaBb	AABB	1/16
AaBbCc	AABBCC	1/64

In general, the probability will be $(1/2)^n$ where n is the number of alleles involved. If you had trouble filling in this table, review frame 16.

27. Speculate on how this information could be used to predict or estimate the number of gene pairs that determine a polygenic trait.

We can not use this approach with individual families, but if large numbers of families can be located in which both parents are known to be heterozygous, we could determine the fraction of offspring with all homozygous genes for the trait in question. Then we would try to match the results with those predicted for different numbers of gene pairs (cf. previous problem). If we knew that the parents were heterozygous, we could study the shape of bar graphs of phenotypes as in frames 14 and 18 to establish whether few or many genes were involved. We would look to see if the curve was steep and if it was continuous or steplike (cf. frame 24). In reality, it is sometimes very difficult to establish the number of genes that determine a quantitative trait. It would be hard to establish that the parents are heterozygous for a trait. Second, the external environment of different individuals is likely to vary, thus modifying the phenotypes. Among other reasons, it is unlikely that individual genes contribute equal increments to a polygenic phenotype. Let us consider this last possibility.

28. In the dihybrid cross discussed above (cf. questions 10 and 11), if the allele H contributed 6 rather than three units to height but all the other alleles (i.e., h, T, t) behaved as above, how would the distribution of phenotypes be affected?

GENOTYPE	FREQUENCY	PHENOTYPE
HHTT		
HhTT		
HHTt		
HhTt		
hhTT		
HHtt		
Hhtt		
hhTt		
hhtt		

GENOTYPE	FREQUENCY	PHENOTYPE
HHTT	1	18
HhTT	2	13
HHTt	2	16
HhTt	4	11
hhTT	1	8
HHtt	1	14
Hhtt	2	9
hhTt	2	6
hhtt	1	4

In summary, the distribution of phenotypes becomes more varied. If the bar graph were drawn, it would be much more spread out and irregular and would be difficult to recognize as being the result of two pairs of alleles.

29. Human skin color is determined by approximately five pairs of alleles that contribute to the production of a brown pigment, melanin. Thus, skin color provides another example of ____________ or ____________ inheritance.

---

polygenic
quantitative

30. Consider a hypothetical couple both of whom are genetically heterozygous with regard to the five pairs of alleles that determine skin color. Assume that both sets of their parents are homozygous and that pigment production is dominant. List the genotypes of the parents and the couple in question.

____

____

____

All parents are AABBCCDDEE and aabbccddee. The couple must be AaBbCcDdEe. (Other parental combinations are also possible such as AABBccDDee and aabbCCddEE.)

31. The couple in the previous question would be dark skinned (but not very dark). Is it theoretically possible for them to have a child as light as the grandparent with all recessive alleles? ________ What is the probability of such an event? ____

Yes

The chance of the child receiving the genes aa from parents that are Aa is 1/4. Similarly, the chance of receiving any other pair of recessive genes is 1/4 and there are five genes pairs. The probability of receiving all recessive alleles is $(1/4)^5$ or 1/1024. (Cf. frame 26A.)

32. Without working out the Punnett square, what can you predict about the shape of the distribution curve for the progeny of this couple?

The curve should be continuous and steep relative to the dihybrid and trihybrid crosses plotted in previous questions (cf. frame 24).

33. Breeders of plants and domestic animals are especially concerned with polygenic traits. Speculate on why this is true.

Financial or practical reasons. Size and weight are generally polygenic traits, and, of course, these two characteristics are of great agricultural importance (e.g., size of eggs, amount of milk produced by cows, and size of pigs).

CHAPTER 10

Environment and Heredity: The Twin Approach

Chapter 9 stressed the impact of environment on quantitative traits. The term "environment" is used by geneticists in a very broad sense to include all the extrinsic forces that interact with living organisms and affect development. Included in this concept would be light, temperature, humidity, altitude, nutrition, disease, and many other factors.

The environment may be assumed to interact with genes in determining every phenotype. Often the degree of this interaction is difficult to assess, especially in humans, because the environment can be neither eliminated nor precisely controlled. Studies of identical and fraternal twins provide one approach to estimating the contribution of both heredity and environment to the expression of a trait.

1. A trait such as height is determined both by ________________ and by ________________. Twins have been studied in attempts to assess the relative contribution of these two factors to height and to dozens of other phenotypes.

- -

heredity (or genes or genotype)
environment

2. Identical twins are more properly called "monozygotic" (MZ) twins because they are derived from a single zygote. That is, both individuals

stem from the same egg and sperm cells. They are, therefore, genotypically identical. MZ twins are always of the same sex. Explain. ____________

____

Since they are both derived from the same sperm and egg cells, they must have received the same sex chromosomes.

3. Fraternal twins are more correctly called dizygotic (DZ) twins. They result from two zygotes (i.e., two different egg cells fertilized by two different sperm). They are, therefore, genotypically (the same, different) . Compare their genetic similarity with a pair of siblings that are not twins.

____

____

---

different

A pair of DZ twins are no more alike genetically than a pair of siblings.

3A. DZ twins may be of the same or opposite sexes. Explain.

____

____

DZ twins are derived from different gametes. Therefore, the sex chromosomes they receive are determined by chance.

Consider the following pedigree charts in the next few questions.

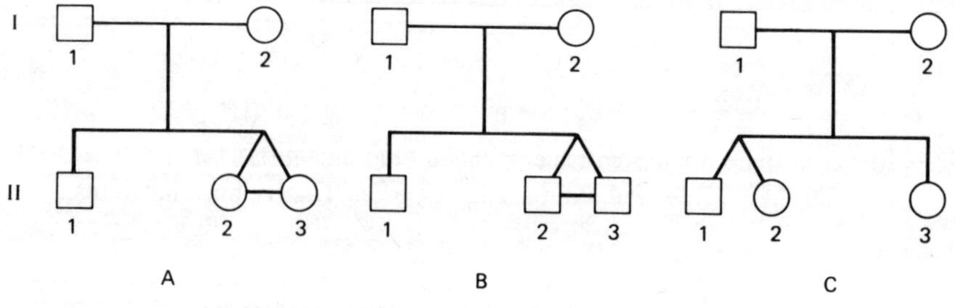

4. Siblings II-2 and II-3 of chart A are ____________ twins whereas II-1 and II-2 of chart C are ____________. Individuals II-2 and II-3 of chart B are genotypically (identical, dissimilar) since they are ____________ twins.

MZ (identical); the clue is in the use of symbols (see chapter III, Introduction).
DZ twins
identical
MZ

5. If II-2 (chart A) married II-2 (chart B), and II-3 (chart A) married II-3 (chart B), and each couple had a child, the relationship between the two children would be ____________. How similar would the genotypes of these two cousins be? ____________
Explain. ____________

---

cousins (first cousins)
genotypically like siblings
The two mothers have identical genes and the two fathers have identical genes.

6. Suppose II-2 (chart A) married II-2 (chart B) as in the previous frame, and II-3 (chart A) married II-1 (Chart C), and each couple had a son. What would be the relationship between the two children? ____________
How similar would these children be genetically? ____________

---

cousins (first cousins)
They would be as genetically similar as stepbrothers since the mothers (II-2 and II-3) are genetically identical.

7. Hereditary enamel hypoplasia is an X-linked dominant trait. Let's look at chart A. Assume that I-1 has this trait and that I-2 does not have enamel hypoplasia. What can we predict about individuals II-2 and II-3?

Both must have the trait.

♀ \ ♂	X^E	Y
X^e	X^EX^e	X^eY

7A. If this trait were autosomal and dominant, what could we predict about individuals II-1, II-2, and II-3, assuming that I-1 was heterozygous?

II-1 and II-2 each have a 50% chance of having the trait since this is a test cross. II-3 should be the same as II-2 since they are genotypically identical.

8. Now refer to chart C. If I-1 is heterozygous for an autosomal dominant trait and I-2 does not have the trait, what can we predict about their offspring?

In each case, the likelihood of having the trait is 50%. Since the twins, II-1 and II-2, are dizygotic, their genotypes and phenotypes are as independent of each other as in nontwin siblings.

9. In chart A, I-1 and I-2 are both tall and homozygous for the genes that establish height. What can we predict about the heights of II-1, II-2, and II-3?

We might expect that all of the progeny would be tall (but read on).

9A. Suppose that, at maturity, II-1 is three inches shorter than his parents, how might we interpret this? ______________________________

____

- -

One possibility is inferior environment.

9B. Give one or two examples of how environment could prevent II-1 from attaining his genetic potential in height. ______________________________

____

____

____

- -

There are an infinite number of possible answers involving such factors as diet, disease, exercise, and prenatal environment.

10. A difficulty with the above analysis is that it would be hard to prove that the two parents were indeed homozygous for all the alleles involved in this polygenic trait. (A few heterozygous pairs of alleles would lead to a range of height among the progeny.) Suppose that II-2 and II-3 (chart A) differ at maturity by three inches; what would this suggest about environment?

____

____

____

- -

The difference in height would suggest an environmental contribution to height since the twins are genetically the same.

11. If twins II-1 and II-2 (chart C) were found to have a different height at maturity, would this prove that the difference is environmentally caused? ____________ Explain. ______________________________

____

____

- -

No. The difference may also be due to a difference in genotypes since these dizygotic twins are not genetically identical. (In fact, we would expect the twins to be of different height since they are of different sexes, and males, in general, are taller than females.)

12. Suppose that we measured the height of 100 pairs of MZ twins and found that, on the average, they differed by less than one inch. What might we infer from this about the contribution of environment to height? _______

____

Does this prove that the environment has little or no effect on height in general? ____

____

Environment had little effect (or an equal effect) on height in each pair of twins. No. One might reasonably argue that the environments of each pair of twins was very similar.

13. A term, "co-twin," is used by geneticists to designate each member of a pair of twins. To solve the problem of the contribution of environment to height, we might compare monozygous co-twins reared under markedly different environments. How could we do this? _______________

____

____

Compare the heights of MZ co-twins who were reared apart from a very young age.

14. Assume that you measured 25 MZ pairs of twins who were raised separately and that they averaged about one inch difference in height. What would this tell us about the relative contribution of environment to height?

____

____

____

It would suggest that the environmental contribution to height is practically nil, and, by deduction, height is determined genetically.

15. If, on the other hand, 25 MZ pairs of twins raised separately were found to average six inches of difference in height, what would your inference be?

____

In this hypothetical case, only 25 pairs of twins were used. Wouldn't 100 pairs of MZ twins (as in frame 12) have been more suitable?

____

Environment does make a significant contribution to the determination of height.
Of course, we would be more comfortable with the statistics if 100 pairs were used, but the likelihood of finding large numbers of MZ twins raised in separate homes is not great.

16. A hypothetical twin study reveals that in 50 MZ twins and 50 DZ twins (of the same sex), the average height difference at maturity is 0.7 inches and 3.3 inches, respectively. How could this be interpreted?

____
____
____
____
____

If we assume that each DZ twin develops in an environment as similar to his co-twin as does each MZ twin to his co-twin, then we might conclude that the genotype makes a heavy contribution to height since the genetically different (DZ) twins are so much more different in height than the MZ twins. (However, it could be that MZ twins have a more similar environment than DZ twins.)

In the "twin method" for estimating the relative degree of contribution of genes and environment to a phenotype, the geneticist determines whether a trait occurs in both co-twins (concordant) or in one twin but not the other (discordant). Concordance in MZ twins is an index of genetic contribution. In the following questions, consider the table below of hypothetical (but reasonable) data.

TRAIT	NUMBER OF PAIRS OF MZ TWINS STUDIED	PERCENT CONCORDANCE
Blood type (ABO series)	100	100
Measles	100	90
Diabetes	100	80
Encephalitis	50	6

17. In the above table, "100% concordance" for blood type means that, in each of the 100 pairs of twins studied, if one twin has a particular blood type (e.g., type B blood), the co-twin has the same blood type. In the three disease studies, at least one co-twin in each pair had the disease. What then is the meaning of 6% concordance for encephalitis?

A group of 50 pairs of twins was studied in which at least one twin had encephalitis. In this group, only 6% of the pairs were concordant for the disease (i.e., in 3 of the 50 pairs, both co-twins had encephalitis). The other 47 pairs (94% of the pairs) were discordant for encephalitis; that is, one of the pair was normal.

18. What are the implications of the concordance data for encephalitis?

Encephalitis is predominantly environmentally caused since the percentage concordance is so low (i.e., 6%). This low value may indicate a very slight genetic predisposition to this disease. But read on.

19. In the table, only half as many twin pairs with encephalitis were included as those with measles. Although these are hypothetical data, they typify the literature. Can you guess why? ______________________________

____

Some traits such as encephalitis are much rarer, and fewer twins with such a trait can be located. This is one of the limitations of the "twin" method.

20. As a <u>first approximation</u>, what would be the interpretation of the data on blood types, measles, and diabetes? ______________________________

____

The high percentage concordance suggests that these traits may be mainly genetically determined.

21. Can you suggest any reasons why we might be skeptical of this tentative analysis? ______________________________

____

____

____

Perhaps this question is too vague or subtle, but our analysis would be tempered if we are dealing with an infectious disease such as measles or if we are concerned with a common childhood disease.

22. How could we experimentally control for the possibility of a twin transmitting a disease to a co-twin? ______________________________

____

____

The best way is to compare the data in the table with a comparable study of MZ twins reared apart. We might also compare the tabulated data with comparable data for DZ twins.

23. Now let's add more data to the above table. Suppose that a study of MZ twins reared apart revealed concordances of 100, 70, and 20% for blood type, diabetes, and measles, respectively. How would this information

affect our tentative interpretation in frame 20? ______

Blood type: since different environments have no effect on blood type, our suspicion that this is genetically determined is confirmed.
Diabetes: environment makes a small contribution, genes a larger one.
Measles: the environment seems to be very important in determining this disease. This is reasonable since we know it is contagious.

24. We may also check the tentative inferences of question #20 by studying a group of DZ twins. Suppose that a study of DZ twins (of the same sex) reveals concordances of 85, 40, and 15% for measles, blood type, and diabetes, respectively. How would these data affect our comments in frame #23? ______

All three figures are consistent with the answer to question #23.
1. Blood type. The high concordance in blood type for MZ twins suggested that genes were important in determination of blood type (frame 20). The 100% concordance for MZ twins reared apart showed that different environments would not affect blood type, suggesting that genetics was more important than environment in regard to this trait. The finding of only 40% concordance of blood types in DZ twins (compared to 100% in MZ twins) again supports the contention that genes play the major role in establishing the blood type since the DZ twins with different genes but similar environments are so much lower.
2. Diabetes. The interpretation of the data on diabetes is similar to that of blood type. Since the percentage concordance for this disease in MZ twins is less than 100, we may infer that genes are not entirely in control (frame 20). The fact that MZ twins reared apart have a high concordance for diabetes indicates that environment does not have a major effect on the expression of this disease (frame 23). The DZ twins (with similar environments) have a low concordancy for diabetes, again suggesting a major genetic role in determination of this disease.
3. Measles. We first suspected that measles might be genetically determined because of the high concordance of this disease in MZ twins (frame 20). However, when the MZ twins were reared apart (thus increasing differences in environment), the concordancy dropped dramatically, suggesting a major role of environment in getting measles (frame 23). Finally, the high concordance for DZ twins suggested that individuals who are genetically very different but are reared together are prone to get the disease, supporting the fact that environment not genes determines measles (frame 24).

25. Analyze the following hypothetical data (as in questions 19-24 above).

TRAIT	PERCENTAGE CONCORDANCE IN TWINS		
	MZ REARED TOGETHER	MZ REARED APART	DZ* REARED TOGETHER
Scarlet fever	70	55	45
Phenylketonuria	100	100	25
Eye color	100	100	30

*Twins of the same sex.

Phenylketonuria (PKU)[1] and eye color are genetic traits little affected by the environment. Scarlet fever is affected significantly both by environment and heredity.

26. DZ twins may be of the same or different sex(es). Which of these two possibilities is more likely? ____________________
Why? ____________________

____

- -

They're equally probable.
The probability of two males or two females is one in four for each or 0.5 for the two possibilities. The probability of one male and one female is two in four or 0.5. If the underlying logic is not apparent, review Chapter 4.

27. Suppose that the census of Philadelphia revealed 65 MZ pairs of twins, 173 DZ pairs of twins of mixed sexes, and 122 DZ twin pairs of the same sex. What would be your suspicion about the data?

____

____

____

____

____

- -

Because there are so many more DZ twins of mixed sexes than of the same sex, one could suggest that some DZ twins of the same sex were mistaken for MZ twins and the 65 MZ twins include a sizable fraction of DZ twins.

In reality, it is not always easy for an observer to distinguish the zygosity of twins (it is especially difficult at birth). The experienced geneticist uses a large series of concordancy tests to establish the zygosity of twins. The geneticist will use a few dozen phenotypic observations of highly penetrant, well-understood traits. Serum protein analysis and blood types are especially useful. Examination of the fetal membranes may also offer useful clues.

NOTES

[1] PKU is a disease involving severe mental retardation; it is characterized by the buildup of compounds called phenylketones in the blood and urine, hence the name.

CHAPTER 11

Multiple Allelism (The Blood Groups of Man)

Although the genes are relatively stable entities, they do change occasionally (a fact that is essential for evolution to occur). Such changes or "mutations" introduce new potentials for all organisms. Frequently, mutations are harmful. Harmful dominant genes will tend to be weeded out of a population by the process of natural selection. Harmful recessive mutations and all mutations that are beneficial or neutral in impact may be sustained. In time, such mutant genes may spread throughout the world's population. If a group of people begins with two varieties of a specific gene (i.e., dominant and recessive), and then mutant alleles are introduced, three or more alternative forms of a gene could result. This phenomenon is well known and is called "multiple allelism." Multiple allelism is exemplified by blood types in man.

1. Although genes tend to be stable and to be copied faithfully from one generation to the next, occasionally a gene may change or ____________. Such changes may be of positive, negative, or neutral value to the host. A mutation introduces new genetic information to an individual (and to the population).

- -

mutate

2. Mutations are normally <u>(rare, frequent)</u> events. When they have no apparent cause, they are called spontaneous mutations. Certain forms of

radiation and some chemicals increase the likelihood of a mutation (i.e., induced mutations). Gene changes due to exposure to X-rays are ________________________________.

---

rare
induced mutations

3. Inasmuch as genes are present in all kinds of cells, all kinds of cells may undergo ______________ or ______________ mutations. This would include the gametes and their precursors (i.e., germ cells) and all other cell types (i.e., somatic cells).

---

induced
spontaneous

4. An example of a somatic mutation would be a change in skin tissue resulting in patches of unpigmented skin. The mutated genes (would, would not) be transmitted to later generations. In order for a mutation to be passed from one generation to the next, it must occur in a ________ cell.

---

would not
germ (Keep in mind that the germ cells are precursors to the gametes and the gametes are the hereditary link between generations carrying the genes from parents to offspring (cf. Chapter 1).

5. Assume a mutation occurs in gene X in a gamete and that this mutant gene is spread throughout a population. How will this affect the number of forms of gene X in that population?

____

____

(Include any necessary qualifications.)

____

____

____

If we had started with two forms of the gene, we would now have three. This assumes that the mutated gene is neither a strong negative nor a strong positive factor in natural selection. (In other words, the assumption is that the new allele does not make the bearer significantly more likely or less likely to propagate.)
Additional mutations to gene X (or the modified gene) will lead to a population containing several varieties of that gene. This is sometimes called polymorphism, a situation in which there are several forms of a species in such proportions that the least common species cannot be explained solely by recurrent mutations.

6. The human blood protein, hemoglobin, is under the control of two pairs of genes. Yet, we can demonstrate numerous inheritable varieties of hemoglobin. These varieties exemplify multiple allelism. The blood types in the so-called ABO series involve several possible varieties of a single pair of genes so they also correspond to ________________.

- -

multiple allelism

To understand the principles underlying blood type (or blood group) inheritance, we must now define some terms. An antigen elicits an immunological response in organisms lacking that substance. The antigen is generally a large complex molecule such as a protein. It may enter the host organism as part of an infectious virus or microbe. The "immunological response" entails production and release of antibodies that are highly specific proteins. These antibodies specifically bind to and tie up or inactivate the antigen. The immune system plays an important role as a defense mechanism against pathogens.

Individuals with Type A or B blood inherit a capacity to synthesize antigenic substances A and B, respectively. Type O individuals synthesize neither antigen. Blood types correspond to phenotypes. These phenotypes may be established by testing a drop of an individual's blood with standard antiserum. Each individual has a pair of alleles for antigen production in the ABO system. Each allele may be one of three kinds (i.e., I^A, I^O, I^B). Actually, there are even more possible allelic forms in the ABO series, but the assumption of three possibilities approximates reality and makes it easy to continue.

7. List the possible genotypes that govern the ABO series

____

- -

I^AI^A, I^AI^O, I^AI^B, I^BI^B, I^BI^O, I^OI^O

8. A person who is genotypically $I^O I^O$ would be phenotypically blood type O. He (would, would not) produce antigen A and (would, would not) produce antigen B.

---

would not
would not

9. Consider only the ABO blood groups. (We will treat the Rh problem later.) Could the individual with Type O blood be safely transfused with Type O serum? ____________________
(A safe transfusion requires the absence of an immunological response that could lead to clumping or agglutination of blood cells.)

Yes. Type O blood contains no antigen. Therefore, it cannot elicit an immunological response.

10. What would be the result of giving Type A or Type B blood to the individual with Type O blood? ____________________
__________Why?____________________
____

---

immunological response (clumping, agglutination)
These blood types contain antigens that he lacks and, therefore, he reacts to them.

11. If a person has Type O blood, what kind of blood may be safely accepted in a transfusion?____________________To what kinds of individuals may he donate his blood?____________________
Explain. ____________________
____

---

only Type O blood
all kinds
Since he produces no antigen in the ABO series, individuals of all blood types will accept his blood without clumping.

12. A person who is genotypically I^AI^B has Type____ blood. He (would, would not) produce antigen A and (would, would not) produce antigen B.

AB
would
would

13. To what types of individuals can he donate his blood?__________ Explain. __________

AB type only
The presence of antigen A prevents transfusion of blood to people with Type O or B blood. The presence of antigen B prevents transfusion to Type O or A individuals.

14. From what blood groups may a Type AB person receive blood?

____

All types. (Types O, A, B, and AB.)
Since he has antigen A and B he will not reject blood containing either or both of these antigens nor will he reject Type O blood which lacks antigen A and B.

15. An individual with Type A blood must be genotypically ________ or ________. He produces antigen________, accepts blood of Type(s) ________, and rejects blood of Type(s) ________.

I^AI^A
I^AI^O
A
O or A
B or AB

16. An individual with Type O blood produces gametes that are genotypically ________.

I^O

17. An individual of Type A blood produces gametes that are ______ or ______ genotypically. A Type AB male produces sperm of genotypes ______ or ______ in a ________ ratio.

I^A or I^O because he may be homozygous (I^AI^A) or heterozygous (I^AI^O)
A
B
1:1

18. What can we say about dominance with genes I^A and I^O ______________________________, I^B and I^O ______________________________, I^A and I^B ______________________________?

I^A is dominant over I^O, I^B is dominant over I^O, neither I^A nor I^B is dominant over the other. (A and B are said to be codominant.)

For frames 19 to 23, refer to the pedigree chart below. Assume that I-1 is Type AB, I-2 and II-4 are Type O, I-3 and I-4 are Type A, and II-3 is Type B.

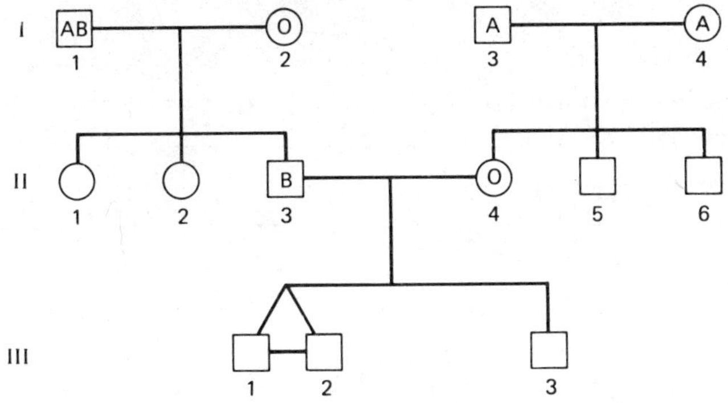

19. What can be deduced about the genotypes and phenotypes of II-1 and II-2?

---

they are I^AI^O (Type A) or I^BI^O (Type B)

20. What is the genotype of individual II-4? ________ What are the genotypes of individuals I-3 and I-4? ________ Explain. ________

I^OI^O
both are I^AI^O
They each must have at least one gene I^A to explain their blood type and at least one gene I^O to give to offspring II-4.

21. What can be deduced about the genotypes and phenotypes of II-5 and II-6?

---

they are I^AI^A or I^AI^O (Type A) or I^OI^O (Type O)

22. What can be stated about the genotype of II-3? ________

I^BI^O

23. What can be determined about the genotypes and phenotypes of III-1, III-2, and III-3? Explain. ______________________________

III-1 and III-2 must be identical in genotype and phenotype since they are MZ twins. III-1, III-2, and III-3 may be I^OI^O (Type O) or I^BI^O (Type B).

In addition to the ABO blood groups, the Rh factor is of great concern in blood typing. The latter antigen is found in about 85% of the American population, and individuals who have this Rh factor or antigen are Rh positive (all others are Rh negative). Immunological incompatibility involving the Rh factor sometimes leads to severe medical complications (i.e., erythroblastosis fetalis, which entails breakdown of red blood cells in the fetus and consequent anemia). Treatment of this disease is possible but requires awareness and preparation. This requires genetic prediction.

24. An individual may be Rh+ or Rh– depending on whether or not he possesses the ______________________ .

Rh factor or Rh antigen

25. Which is the more likely possibility?

Rh+ (85% of the population)

26. An individual who is Rh+ has either one or two genes for production of ________________. The Rh– person must be homozygous for the genes corresponding to absence of this antigen. The gene for antigen production is (dominant, recessive) .

---

antigen
dominant

27. List the possible genotypes for an Rh+ person and for an Rh– person.

____

Rh+: RR or Rr
Rh–: rr

28. It has been demonstrated that although we are dealing with one pair of genes, distinguishable varieties of inheritable Rh factors are found in the population. In other words, the antigens of Rh+ individuals may be of several kinds. This, once again, exemplifies ______________________ ______________________ .

---

multiple allelism (or polygenic inheritance or polymorphism)
Blood Types A and B are also polymorphic.

29. An Rr individual is phenotypically______. He would produce gametes of genotypes R and r in a ratio of______. What possible genotypes could his parents have been?______________________ ______________________

Rh+
1:1
RR x Rr or RR x rr or Rr x Rr or Rr x rr

30. In an Rh– person, what can we state about his genotype?________, the genotype of his gametes?______, and the possible genotypes and phenotypes of his parents?______________________ ______________________

---

rr
r
Rr x Rr (both Rh+) or Rr x rr (one Rh+ and one Rh–) or rr x rr (both Rh–)

31. The disease erythroblastosis fetalis results from an immunological reaction of the mother's Rh– blood with the Rh antigen. Consider a fetus derived from Rh+ parents. What would be the possible genotype(s) and phenotype(s) for that individual? ______

______

RR or Rr (Rh+) or rr (Rh–). His parents could be homozygous or heterozygous.

32. Is erythroblastosis fetalis likely to occur in this instance?______. (Note: this disease entails blood group incompatibility and works like ABO incompatibility. Cf. frames 9 to 18.)

No. Even if the fetus is Rh+, the factor will not be treated as an antigen by the mother since it is not a "foreign substance."

33. If both parents are Rh–, what can be stated about the genotype(s)______ and phenotype(s)______of their progeny? Is there any likelihood of an immunological reaction?______ Explain.______

______

---

rr

Rh–

no

The fetus does not produce the Rh factor so there can be no immunological response.

34. If the father is Rh– and the mother is heterozygous, specify the possible genotype(s) and phenotype(s) of the fetus.______

Comment on the likelihood of immunological incompatibility between mother and fetus. ______

______

______

______

Rr (Rh+) or rr (Rh–)
No incompatibility. If the child is Rr and produces Rh factor, the mother's blood would not "recognize" the factor as "foreign" since she is also Rh+. Thus, no immunological reaction could occur.

35. What would be the possible genotype(s) and phenotype(s) from a mating of an Rh+ father and an Rh– mother? ______
Is it possible for erythroblastosis fetalis to occur? ______ Is this outcome inevitable? ______ Explain. ______

Rr (Rh+) or rr (Rh–)
yes
no
If the child is Rh–, there is no chance for an immunological reaction. (Even if the child is Rh+, the pregnancy may be free of problems. For example, the Rh factor may not cross the placental barrier to his mother's blood.) In practice, the likelihood of immunological problems increases with later pregnancies. A woman may become sensitized to the Rh antigen and react more dramatically.

36. Blood typing is sometimes applied to the determination of zygosity of twins (cf. Chapter 10). Given only the following data, what can be stated about possible pairs of twins?

INDIVIDUAL	BLOOD GROUP
A	B, Rh–
B	AB, Rh–
C	O, Rh+
D	B, Rh–
E	B, Rh+

______

Individuals A and D (but no other pairs) may be monozygotic twins although it is possible that their phenotypic identity is coincidental. Any other combination (e.g., A and B, C and E) could conceivably be dizygotic twins. (In practice, additional traits and other blood antigens are required for establishment of zygosity as discussed in Chapter 10, Frame 27.)

37. Anthropologists use blood group data to establish the kinship of different cultures. For example, from the following hypothetical data (which approximate the values in the literature), what may be inferred about the origin of gypsies?

POPULATION	PERCENTAGE			
	0	A	B	AB
Hungarians	33	38	20	9
Hungarian gypsies	32	18	41	9
Indians (East India)	30	16	44	10

These data (along with other information) support the contention that Hungarian gypsies have migrated from India. At any rate, they are genetically very different from other Hungarians.

38. A final and not uncommon application of blood group data is to help determine paternity in disputed cases. For example, if a mother is O– (i.e., Type O, Rh–) and her son A–, could Mr. X, Mr. Y, or Mr. Z be the father? See below:

Mr. X: B–
Mr. Y: AB+
Mr. Z: O+ ________________

X or Z: no; Y: yes

39. If your answer was "Yes," then can we determine paternity absolutely from these data? ________________

No. We can definitely rule out Mr. X and Z, but the data constitute support rather than positive proof that Y is the father.

FINAL EXAMINATION

Having faithfully studied all the frames in this book and answered all of the questions, you may want some feedback on your overall retention. The purpose of this examination is for you to test yourself for mastery of the various aspects of human heredity.

I. In man, a slight downward pointed bulge of skin on the inner rim of the ear is known as Darwin's ear point. It is controlled by an autosomal gene (i.e., not X-linked). In addition, ear lobes can be attached or free (hanging), and this trait is controlled by a different pair of alleles on another chromosomes (also an autosome). From the pedigree chart below and Table 1:

(a) Establish with unambiguous arguments which alleles are dominant and which are recessive. (For example, is Darwin's ear point dominant or recessive? Why?)

(b) Deduce as many genotypes as possible. (Not every genotype can be established with the data available.)

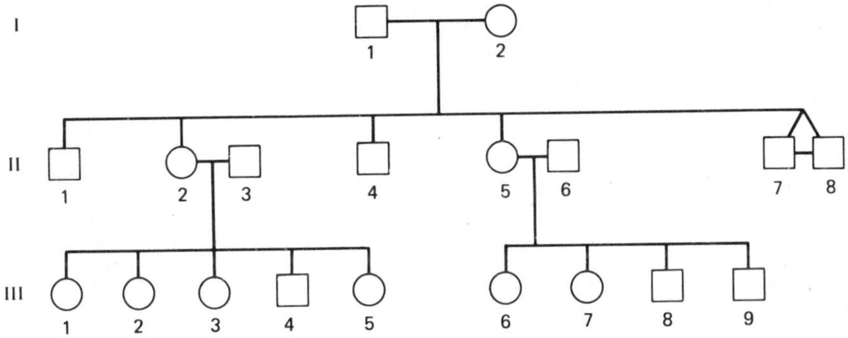

TABLE 1

INDIVIDUALS	PHENOTYPE
I-1, II-2, 7, 8, III-4, 5, 7, 9	Absence of ear point, attached ear lobes
I-2, II-5, 6	Presence of ear point, attached ear lobes
II-3, III-1, 2, 3	Absence of ear point, free ear lobes
II-1, 4, III-8	Presence of ear point, free ear lobes
III-6	Presence of ear point, unknown ear lobes

Possible Approach (Clues)

1. First consider each trait by itself. Is "free" or "attached" ear lobe dominant? Is presence or absence of ear point dominant?
2. Define your symbols.

II. This problem is also based on the above pedigree. We are going to add new complications to the above conditions to see how each new variable or condition changes the resulting pedigree.

Case 1

Assume that free ear lobes is an X-linked trait but that Darwin's ear point is autosomal dominant. Consider only II-2 and II-3 and their offspring (III-1, 2, 3, 4, 5). From the data in Table 1, deduce the genotypes of the offspring.

Case 2

Now we'll consider a different possibility (and forget the case 1 assumptions). Assume now that II-5 and II-6 are heterozygous for both traits and that the genes for Darwin's ear point (dominant) and for free ear lobes (dominant) are linked very closely to the same autosome. How would the linkage affect the genotypes of III-6, 7, 8, 9? Are the phenotypes shown in Table 1 for the couple (II-5 and II-6) and their offspring possible if linkage is assumed? Explain.

Case 3

Now, we'll assume that attached ear lobes (dominant) are determined by three pairs of alleles (polygenic) instead of just one pair. If you assume that I-1 and I-2 are heterozygous with respect to all three pairs of alleles for ear lobe types, then what would you predict about individuals II-1, 2, 4, 5, 7, 8? Explain. (Consider only the ear lobe trait and the individuals cited.)

Case 4

Monozygous twins II-7 and II-8 both have dimples on their right cheeks, both have hairy chests, and both have an allergy to genetics exams (they get rashes). Twin II-7 is overweight; twin II-8 is underweight. What can be inferred about the possible contributions of genes and environment to these traits? Explain.

III. We will assume, for this next problem, that there is a specific trait ("thick skull") that is rare and inheritable as shown below. (Shaded circles and squares have this trait.) (a) Determine if thick skulls are inherited by any of the common modes of inheritance. (b) Explain logically and thoroughly how you came to your conclusion. (c) If this trait were not so rare, would your interpretation differ? Explain.

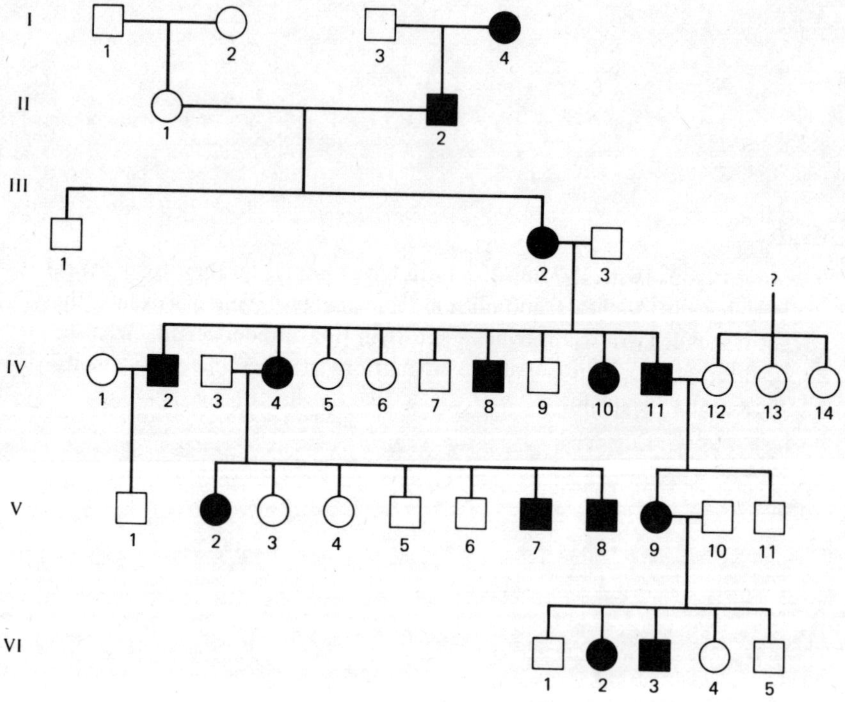

IV. You have just been selected to be a member of a panel of experts who will judge in the following case. Woman X is deceased; she was the mother of four children. Man Y and Man Z both claim to have fathered these children. Assuming that one of these two men is the father of all four, assign the children to their lawful father. The blood types are given below. Explain your answer.

Woman X:	Blood Type A, Rh–
Man Y:	Blood Type B, Rh+
Man Z:	Blood Type AB, Rh+
Child 1:	Blood Type AB, Rh–
Child 2:	Blood Type B, Rh+
Child 3:	Blood Type O, Rh+
Child 4:	Blood Type AB, Rh–

V. Assume that the pedigree chart below is for a rare autosomal trait. (a) Is the gene for this trait recessive or dominant? (b) Fill in as many genotypes as possible for the individuals shown in the chart. (c) What is the likelihood for the distribution of sexes in the offspring (i.e., three males and two females)? (d) If the couple had one more child, what is the probability that the child would have the trait? (e) If the couple had two more children, what is the probability that there would have been two normal daughters?

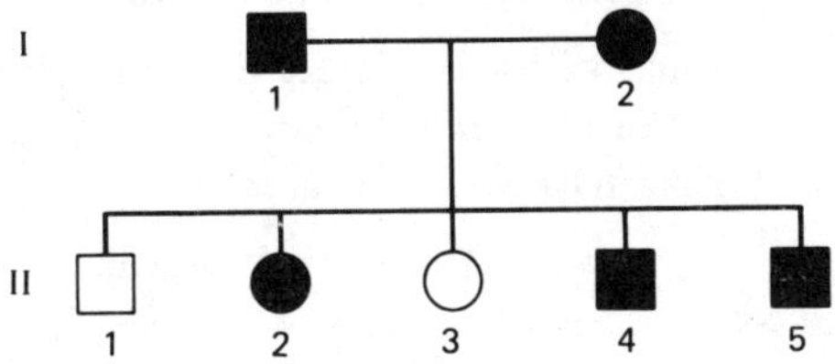

ANSWERS TO FINAL EXAMINATION

I.

(a) 1. Darwin's ear point cannot be recessive. We can refute this possibility by studying the matings of II-5 and II-6. Both individuals have Darwin's ear point. If the gene for this trait were recessive, then II-5 and II-6 would have to be dd and their offspring would *all* be dd and have Darwin's ear point. However, we note that III-7 and III-9 do not have an ear point, therefore, the trait cannot be recessive.

2. The same approach may be used to refute the possibility that attached ear lobes is recessive since II-5 and II-6 both have attached ear lobes and not all of their children have attached ear lobes.

3. Since Darwin's ear point and attached ear lobes can only be recessive or dominant, and each has been shown not to be recessive, we would expect that they are dominant traits. Careful study of the chart shows that the dominant hypothesis cannot be negated.

(b) 1. Assign symbols: D = Darwin's ear point
d = absence of ear point
A = attached ear lobes
a = free (or hanging) ear lobes

2. Considering only Darwin's ear point, we can say that the following are dd since they have the recessive condition: I-1, II-2, 7, 8, III-4, 5, 7, 9 and II-3, III-1, 2, 3. Based solely on the knowledge that Darwin's ear point is dominant, we can designate the following as D– (i.e., one allele for the dominant trait and one undetermined allele): I-2, II-5, 6 and II-1, 4, III-8 and III-6.

3. Relating only to Darwin's ear point, the total information may be presented as a revised pedigree chart.

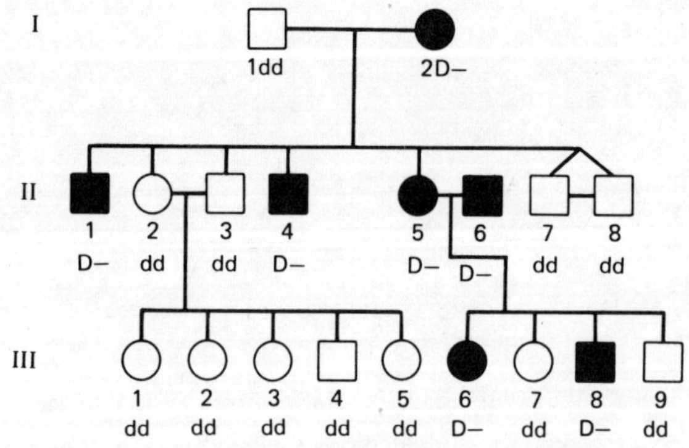

4. Since I-2 has children without Darwin's ear point, she would be Dd. Similarly, II-5 and II-6 cannot be homozygous dominant; they too must be Dd. All offspring of I-1 must have at least one allele d, therefore, II-1 and II-4 must be Dd. Finally III-6 and III-8 are ambiguous–either DD or Dd. We can add the deduced genotypes to the pedigree chart to get the following:

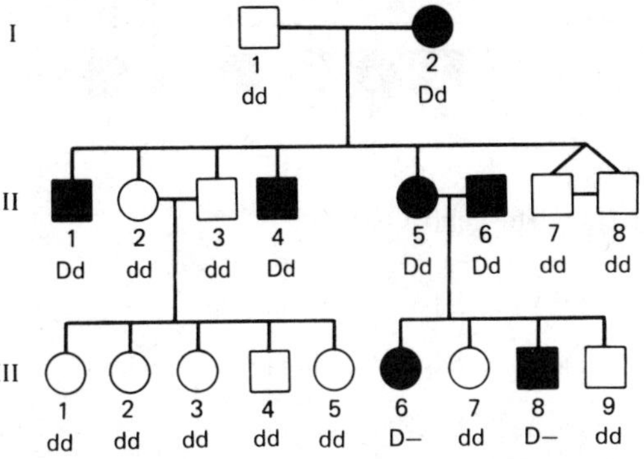

5. Now consider attached ear lobe that we know to be dominant. We may assign "aa" to all individuals with free ear lobes and "A–" to those with attached ear lobes so the pedigree for attached ear lobes looks like this:

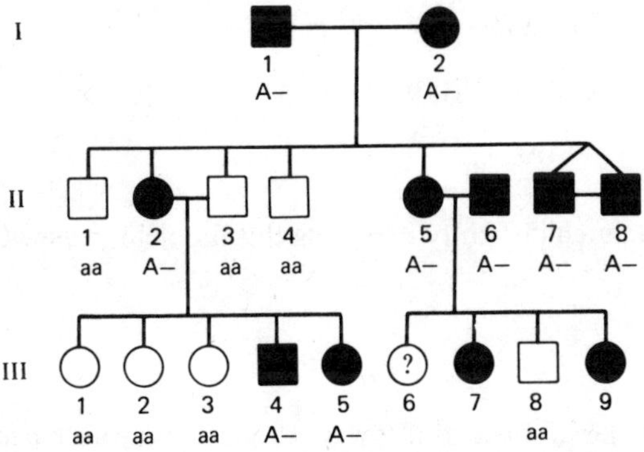

6. Individuals I-1, I-2, II-2, II-5, and II-6 must be heterozgyous since they have at least one child with free ear lobes. Also, III-4 and III-5 must be Aa since their father is aa. All others with attached ear lobes may be AA or Aa.

7. We can now add our deductions to the chart on attached ear lobes as follows:

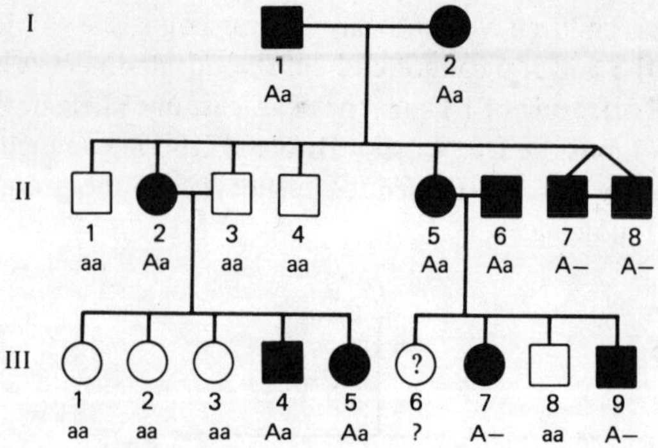

8. In summary then, the genotypes are:

I-1	ddAa	III-1	ddaa
I-2	DdAa	III-2	ddaa
II-1	DdAa	III-3	ddaa
II-2	ddAa	III-4	ddAa
II-3	ddaa	III-5	ddAa
II-4	Ddaa	III-6	D---
II-5	DdAa	III-7	ddA-
II-6	DdAa	III-8	D-aa
II-7	ddA-	III-9	ddA-
II-8	ddA-		

If you had much difficulty in solving this problem, review Chapters 2, 3, and 4.

II.

Case 1

1. We can set up the problem as if it were about two separate pedigree charts and show the key features:

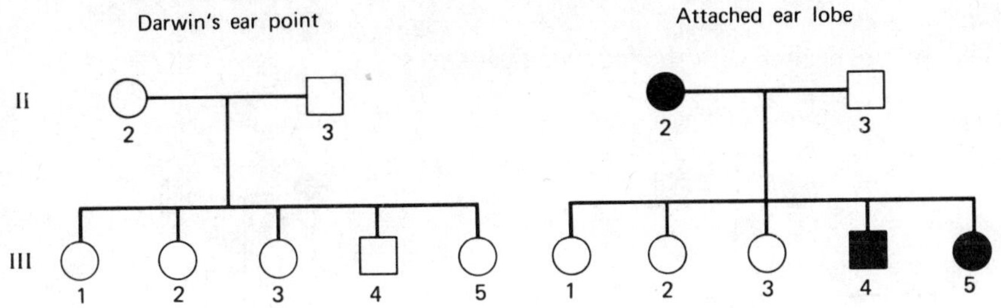

2. Since none of the individuals have Darwin's ear point, they all are homozygous recessive or dd.

3. Attached ear lobe must be dominant to explain the pattern. (If it were recessive, then II-3 would have to have this trait in order to explain III-5.) Therefore, II-2 must be $X^AX^?$; II-3 is X^aY; III-1, 2, 3 and all X^aX^a; III-4 is X^AY; and III-5 is $X^AX^?$.

4. If II-2 were X^AX^A, then all her children would have attached ear lobes. They don't, so II-2 must be X^AX^a. Correspondingly, III-5 must be X^AX^a since she received X^a from her father's sperm.

5. In summary the genotypes are as follows:

II-2	ddX^AX^a (or $\frac{d}{d}$, $\frac{A}{\,}$)
II-3	ddX^aY
III-1	ddX^aX^a
III-2	dd X^aX^a
III-3	dd X^aX^a
III-4	dd X^AY
III-5	dd X^AX^a

6. For a review of the principles involved in this problem, see Chapter 7.

Case 2

1. We are dealing with the following pedigrees:

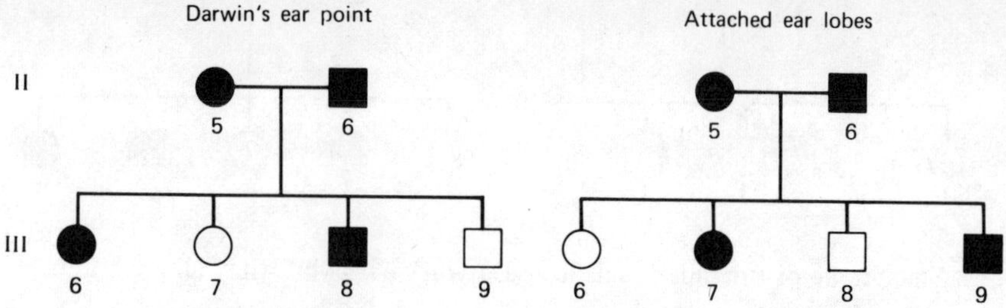

2. If we ignore the linkage momentarily, then II-5 and II-6 would both be DdAa and the phenotypes of all the progeny are feasible.
3. Since there is linkage, the following possibilities exist for II-5 and II-6:

(a) both $\frac{DA}{da}$

(b) both $\frac{Da}{dA}$

(c) one $\frac{DA}{da}$ the other $\frac{Da}{dA}$

4. Since the linkage is very close, crossing over is unlikely.
5. Consider each of the three possible combinations for II-5 and II-6 and the predicted offspring. The Punnett square may be used as shown below. It is seen that combination (a) would lead to only two types of offspring (i.e., both traits or neither trait) and would not explain how III-6, 7, 8, and 9 can have one trait and not the other. Combinations (b) and (c) are possible.

(a)

	$\underline{DA}$	$\underline{da}$
$\underline{DA}$	$\frac{DA}{DA}$	$\frac{DA}{da}$
$\underline{da}$	$\frac{DA}{da}$	$\frac{da}{da}$

Combination contradicted by pedigree chart

(b)

	Da	dA
Da	Da/Da	Da/dA
dA	Da/dA	dA/dA

(c)

	DA	da
Da	DA/Da	Da/da
dA	DA/dA	da/dA

Combinations consistent with pedigree chart

6. For a review of the problems involved in linkage, see Chapter 8.

Case 3

1. We are dealing only with free or attached ear lobes. The pedigree chart below summarizes the data given.

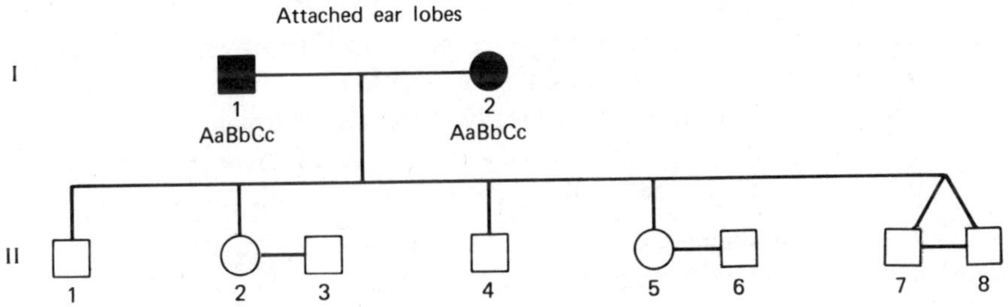

2. Essentially we are dealing with a trihybrid mating in which the genes for ear lobe are additive so they hang to different lengths.

3. The results may be predicted by a Punnett square or by expansion of $[(3X) + (1Y)]^3$ or simply as a bell-shaped curve (if we assume incomplete dominance and equal contribution of each of the three genes). Most of the offspring should have intermediate length ear lobes although there is a slight possibility for either attached or fully hanging ear lobes.

4. For more review on this subject, see Chapter 9 (especially frames 15 to 20).

Case 4

1. MZ twins have identical genotypes so the differences in phenotypes are traceable to environmental causes. Concordance in a single pair of monozygous twins is consistent with a genetic basis but does not prove that the trait is genetic; the similarity could be coincidental.

2. Thus, in view of the limited evidence, the following may have a genetic basis: dimples, hairy chests, allergy to genetics exams.

3. On the other hand, weight is environmentally determined since two genotypically identical twins are so different. (These were meant as hypothetical examples; in reality, both genes and environment influence an individual's weight.)

4. See Chapter 10 if a review of twin studies is desired.

III.

1. Since the trait is determined "by a simple mode of inheritance," we may presume it is autosomal or X-linked and dominant or recessive.

2. Let us consider first the autosomal recessive possibility. It is a possibility *only* if the following individuals are heterozygous: I-3, II-1, III-3, IV-3, IV-12, and V-10. This is extremely unlikely since we are told that the trait is rare. (For example, the probability of a heterozygous carrier of the trait marrying into the family might be 1 in 100; the probability of six such carriers in the pedigree would then be $(1/100)^6$ or 10^{-12} (one chance in a trillion).

3. We may easily refute the X-linked recessive hypothesis by the same arguments as above. Also, we would not expect to see so many females with a rare recessive X-linked trait. Finally, if the trait is X-linked and recessive, all affected females would have affected fathers and this is not the case in the pedigree (cf. IV-2 and VI-2, for examples).

4. Autosomal dominance would seem to be a possibility if each of the following were heterozygous: II-2, III-2, IV-2, IV-4, IV-11, and V-9. Since this is a rare trait, it is very likely that these individuals are heterozygous. Also, each of these has one normal (homozygous recessive) parent. We note also that the ratio of males to females with the trait is 1:1 (i.e. 7 males and 7 females) which is generally more consistent with autosomal than X-linked traits but read on.

5. At this point, the novice would have accepted the autosomal dominant explanation and proceeded to the next question. Hopefully, you checked out the X-linked dominant hypothesis carefully. If the trait is X-linked dominant, then all matings of ●—□ should be equivalent to X^TX^t females x X^tY males. The prediction would be for offspring to be in a ratio of

affected femals : normal females : affected males : normal males of 1:1:1:1. The actual findings for offspring from this kind of mating is four affected females, five normal females, seven affected males, and six normal males and these numbers are reasonably consistent with a 1:1:1:1 ratio considering the small size of the family.

♂ \ ♀	X^T	X^t
X^t	X^TX^t	X^tX^t
Y	X^TY	X^tY

6. The more critical mating for a test of the X-linked dominant hypothesis is ■—○ for which we would predict that *all* sons would be normal and all daughters would be *affected*.

♀ \ ♂	X^T	Y
X^t	X^TX^t	X^tY

Unfortunately, there are only five offspring from such matings but *all five* are consistent with the X-linked dominant hypothesis. These results could be a coincidence but the chance of such results with the autosomal dominant hypothesis is $(1/2)^5$ or only 1 chance in 32.

7. In summary, the trait is very probably X-linked dominant although there is a small probability that the trait is autosomal dominant. The limited data do not permit an absolute conclusion.

8. If the trait were *not* rare, the autosomal recessive hypothesis would become more tenable. The X-linked recessive hypothesis would still be invalid since we note affected women with normal fathers in the pedigree.

9. For more review of this kind of problem, see Chapter 7, especially frames 45 and 46.

IV.

1. We may consider all the combinations of possible genotypes and phenotypes in the offspring.

2. Considering first the woman and man Y, the offspring may be blood Type A or B or AB or O and they may either be Rh positive or Rh negative. In other words, they could produce the combination of children observed. Mr. Y could be the father although the data do not prove he is the father.

Mr. Y

Ms. X ♀ \ ♂	I^B
I^A	I^AI^B

or

Mr. Y

Ms. X ♀ \ ♂	I^B	I^O
I^A	I^AI^B	I^AI^O

or

Mr. Y

Ms. X ♀ \ ♂	I^B
I^A	I^AI^B
I^O	I^BI^O

or

Mr. Y

Ms. X ♀ \ ♂	I^B	I^O
I^A	I^AI^B	I^AI^O
I^O	I^BI^O	I^OI^O

and

Mr. Y

Ms. X ♀ \ ♂	R
r	Rr

or

Mr. Y

Ms. X ♀ \ ♂	R	r
r	Rr	rr

3. Now consider Mr. Z. Mr. Z and woman X could have children who were Type A, Type B, Type AB but *not* Type O. They could have Rh positive or Rh negative offspring as shown above for Mr. Y and Ms. X. We note that child 3 is Type O and could not have Mr. Z for a father. Consequently, the lawful father is Mr. Y.

Mr. Z

Ms. X ♀ \ ♂	I^A	I^B
I^A	I^AI^A	I^AI^B

or

Mr. Z

Ms. X ♀ \ ♂	I^A	I^B
I^A	I^AI^A	I^AI^B
I^O	I^AI^O	I^BI^O

4. For more practice in this kind of problem, see Chapter 11, especially frames 38 and 39.

V.

(a) The trait is dominant. If it were recessive, I-1 and I-2 would be tt and all their offspring would be affected. Individuals II-1 and II-3 eliminate the autosomal recessive possibility.

(b) I-1 and I-2 are Tt (T being the dominant allele for the trait). Note that neither parent could be homozygous dominant or all the offspring would be. II-1 and II-3 are tt. II-2, II-4, and II-5 are TT or Tt.

(c) Expand the binomial $(X + Y)^5$:

$$(X + Y)^5 = 1X^5 + 5X^4Y + 10X^3Y^2 + 10X^2Y^3 + 5XY^4 + 1Y^5$$

If we let X stand for males and Y for females, then the term X^3Y^2 describes the siblings (three males, two females) and the chance of this combination occurring is 10 (the coefficient of X^3Y^2) divided by 32 (the sum of coefficients or 1 + 5 + 10 + 10 + 5 + 1). The answer is 10/32 or 5/16.

(d) P = 0.75 (since we are dealing with a dominant trait in a monohybrid mating).

(e) The probability of a daughter is 1/2. The probability of a normal individual is 1/4. The probability of a normal daughter then is 1/8 (1/2 x 1/4). The probability of two normal daughters is 1/64 (i.e., 1/8 x 1/8).

For more review of this question, see Chapters 2, 3, 6, and 7, and expecially Chapter 6, frames 21 and 22, and Chapter 7, frames 37 and 38.

Bibliography

We hope that you will want to know more about genetics. For further reading, a list of books on human genetics and related topics is given below. This brief list is not intended to be complete, but it provides the reader with an introduction to a wide variety of interesting literature.

Burns, G. W. 1972. *The Science of Genetics*, 2nd Edition. Macmillan, New York.

Court Brown, W. M. 1967. *Human Population Cytogenetics*. North Holland Publishing Co., Amsterdam.

Fuhrmann, W. and F. Vogel. 1969. *Genetic Counseling*. Springer-Verlag, New York.

Gardner, E. J. 1972. *Principles of Genetics*, 4th Edition. John Wiley and Sons, Inc., New York.

Giblett, E. R. 1969. *Genetic Markers in Human Blood*. Blackwell Scientific Publications, Oxford, England.

Goodman, R. M. (Ed.). 1970. *Genetic Disorders of Man*. Little, Brown and Co., Boston.

Harris, H. 1970. *The Principles of Human Biochemical Genetics.* North Holland Publishing Co., Amsterdam.

Herskowitz, I. H. 1973. *Principles of Genetics.* Macmillan, New York.

Hsia, D. 1966. *Inborn Errors of Metabolism, Part 1, Clinical Aspects,* 2nd Edition. Year Book Medical Publishers, Chicago.

Lenz, W. 1963. *Medical Genetics.* E. F. Lanzl, Trans. The University of Chicago Press.

Lerner, I. M. 1968. *Heredity, Evolution and Society.* W. H. Freeman and Co., San Francisco.

Levine, L. 1973. *Biology of the Gene*, 2nd Edition. C. V. Mosby Co., St. Louis.

Levitan, M. and A. Montagu. 1971. *Textbook of Human Genetics*. Oxford University Press, New York.

McKusick, V. A. 1969. *Human Genetics*, 2nd Edition. Prentice-Hall, Inc., Englewood Cliffs, New Jersey.

McKusick, V. A. 1971. *Mendelian Inheritance in Man*, 3rd Edition. Johns Hopkins University Press, Baltimore.

McKusick, V. A. 1972. *Study Guide–Human Genetics*, 2nd Edition. Prentice-Hall, Englewood Cliffs, New Jersey.

Moody, P. A. 1967. *Genetics of Man*. W. W. Norton and Co., New York.

Motulsky, A. G. (Ed.) 1970. *Counseling and Prognosis in Medical Genetics*. Hoeber, New York.

Race, R. R. and R. Sanger. 1968. *Blood Groups in Man*, 5th Edition. Blackwood Scientific Publications, Oxford, England.

Roberts, J. A. F. 1970. *An Introduction to Medical Genetics*, 5th Edition. Oxford University Press, New York.

Singleton, W. R. 1967. *Elementary Genetics*, 2nd Edition. D. Van Nostrand Co., Inc., Princeton, New Jersey.

Srb, A. M., R. D. Owens, and R. S. Edgar. 1965. *General Genetics*, 2nd Edition. W. H. Freeman and Co., San Francisco.

Stern, C. 1973. *Principles of Human Genetics*, 3rd Edition. W. H. Freeman and Co., San Francisco.

Strickberger, M. W. 1968. *Genetics*. The Macmillan Co., New York.

Thompson, J. S. and M. W. Thompson. 1966. *Genetics in Medicine.* W. B. Saunders Co., Phila.

Watson, J. D. 1970. *Molecular Biology of the Gene*, 2nd Edition. W. A. Benjamin, Inc., New York.

Whittinghill, M. 1965. *Human Genetics and Its Foundations*. Reinhold Publishing Corp., New York.

Winchester, A. M. 1966. *Genetics*, 3rd Edition. Houghton Mifflin Co., New York.

Winchester, A. M. 1971. *Human Genetics*. C. E. Merrill Publishing Co., Columbus, Ohio.